Affordance Theory in Game Design

A Guide Toward Understanding Players

Synthesis Lectures on Games and Computational Intelligence

Editor
Daniel Ashlock, *University of Guelph*

Synthesis Lectures on Games & Computational Intelligence is an innovative resource consisting of 75-150 page books on topics pertaining to digital games, including game playing and game solving algorithms; game design techniques; artificial and computational intelligence techniques for game design, play, and analysis; classical game theory in a digital environment, and automatic content generation for games. The scope includes the topics relevant to conferences like IEEE-CIG, AAAI-AIIDE, DIGRA, and FDG conferences as well as the games special sessions of the WCCI and GECCO conferences.

Affordance Theory in Game Design: A Guide Toward Understanding Players
Hamna Aslam and Joseph Alexander Brown
2020

General Video Game Artificial Intelligence
Diego Pérez Liébana, Simon M. Lucas, Raluca D. Gaina, Julian Togelius, Ahmed Khalifa, and Jianlin Liu
2019

On the Study of Human Cooperation via Computer Simulation: Why Existing Computer Models Fail to Tell Us Much of Anything
Garrison W. Greenwood
2019

Exploring Representation in Evolutionary Level Design
Daniel Ashlock
2018

On the Design of Game-Playing Agents
Eun-Youn Kim and Daniel Ashlock
2017

Game Theory: A Classical Introduction, Mathematical Games, and the Tournament
Andrew McEachern
2017

Affordance Theory in Game Design: A Guide Toward Understanding Players

Hamna Aslam and Joseph Alexander Brown

ISBN: 978-3-031-00995-2   paperback
ISBN: 978-3-031-02123-7   ebook
ISBN: 978-3-031-00172-7   hardcover

DOI 10.1007/978-3-031-02123-7

A Publication in the Springer series
SYNTHESIS LECTURES ON GAMES AND COMPUTATIONAL INTELLIGENCE

Lecture #6
Series Editor: Daniel Ashlock, *University of Guelph*
Series ISSN
Print 2573-6485   Electronic 2573-6493

Affordance Theory in Game Design

A Guide Toward Understanding Players

Hamna Aslam and Joseph Alexander Brown
Innopolis University

SYNTHESIS LECTURES ON GAMES AND COMPUTATIONAL INTELLIGENCE #6

ABSTRACT

Games, whether educational or recreational, are meant to be fun. How do we ensure that the game delivers its intent?

The answer to this question is playtesting. However, a haphazard playtest process cannot discover play experience from various dimensions. Players' perceptions, affordances, age, gender, culture, and many more human factors influence play experience. A playtest requires an intensive experimental process and scientific protocols to ensure that the outcomes seen are reliable for the designer.

Playtesting and players' affordances are the focus of this book. This book is not just about the playtest procedures but also demonstrates how they lead to the conclusions obtained when considering data sets. The playtest process or playtest stories differ according to the hypothesis under investigation. We cover examples of playtesting to identify the impact of human factors, such as age and gender, to examine a player's preferences for game objects' design and colors. The book details topics to reflect on possible emotional outcomes of the player at the early stages of game design as well as the methodology for presenting questions to players in such a way as to elicit authentic feedback.

This book is intended mainly for game designers, researchers, and developers. However, it provides a general understanding of affordances and human factors that can be informative for readers working in any domain.

KEYWORDS

game design, human-computing interfaces, affordances, playtesting, user experience

Contents

Preface

This book is primarily written for game designers, although it is not intended to be limited only to them. The reader needs to have an enthusiasm for games and curiosity about players. The reader will understand that player feedback is not just about "I like this game" or "I do not enjoy this game." There is much more hidden behind a player's fondness and dislike for the game.

The book lays out methodologies for player interactions. Methodology of playtest differs depending on the research question and which aspect of design is under investigation. The key to a good playtest methodology is not just a simple gameplay session. The playtester must feel immersed in the game during the playtest. This ensures thorough feedback from the playtesters.

While testing a game, there are not many options available to experiment with the playtest settings. However, by focusing on a specific game aspect rather than a particular game, a playtest setting can be developed to which a playtester can relate. This requires some understanding of the playtester's cultural background, age, work environment, etc. Readers will learn simple and interesting techniques and concepts for creating joyful playtest storylines. The book also gives insight into factors that influence play experience but are ignored.

Referring to the earlier point that the book is not limited to game designers or people having interest in games or game design, anyone who is interested in knowing their audience can benefit from this book. Playtest is a word attributed to game design investigation and improvisation although any interaction where there is an intention of identifying patterns of behavior, certain outcomes, action and reaction trends, etc. falls under the category of the playtest. A teacher standing in the classroom is conducting a playtest that identifies the effectiveness of teaching design. A lawyer, a judge, or a suspect in the courtroom are in a playtest that measures the quality of the jurisdiction system design from different aspects. We are all constantly submerged in a playtest process, consciously or unconsciously. Therefore, we believe, anyone can gain knowledge and learn skills from this book as long as they have the intention of identifying and investigating various aspects of human interactions in a process, with a design, or for a cause.

The book is written with the intention of keeping things simple so that readers do not feel the unnecessary load of heavy words. The chapters are concise and provide concrete details. We, who are proponents of intuitive design and thin rule books, could not practice otherwise. Therefore, the book is furnished with only as much information as needed.

What the readers will learn depends on them and their goals. The purpose of the book is not to teach a definite skill that the readers can use to achieve their goals; rather, it is to equip them with the tools necessary to provide strong support during their endeavors to conquer the mysteries of human mind and worldly magic. Certainly, you as the reader will be able to initiate a process of human discovery and much more. With the passage of time, many new stories will develop that will add to this process, and the process will improve to get us closer to the notion of universal design, because, as we agree, universal design is a process rather than an achievement.

Hamna Aslam and Joseph Alexander Brown
March 2020

Acknowledgments

The authors wish to thank Innopolis University for its support of the Artificial Intelligence in Games Development Lab and the Innopolis Community for engaging and supporting the research in this document. We would especially highlight Giancalro Succi, Sergey Masiagin, and Ekaterina Protsko.

A number of researchers have made comments on this book as well as the source materials and we specifically thank Mike Cook, Matthew Guzdial, Rhoda Ellis, Terry Trowbridge, Daniel Ashlock, Munir Makhmutov, Marco Scirea, and the anonymous reviewers (both for papers accepted and rejected) for their comments and time.

We would also like to thank the co-authors of the research papers which have been used to compile some of the chapters of the book. In this regard, we thank Kamilla Borodina, Ecaterina Baba, Evgenii Nikolaev, and Elizabeth Reading. We thank Abdulkhamid Muminov, Rishat Maksudov, Danila Romanov and Muhammad Ziad Alkabakibi for data collection and cleaning. We are thankful to Rabab Marouf who proofread and provided detailed feedback—of course all remaining errors belong solely to us.

We are thankful to our student and a talented photographer, Maxim Korsunov, for taking pictures for this book and making himself available on short notices. Thanks also go to Lesia Poliakova for her photgraphy. We thank our colleagues and friends in Innopolis, especially the faculty support office for providing a positive environment.

We would like to thank Elizabeth Reading who, beyond being Joseph's most faithful and loving companion, worked diligently with a red pen to help with the proofreading—of course all remaining errors belong solely to us.

Hamna is grateful for her parents' and siblings' support, which kept her spirits up during tiring days.

Hamna Aslam and Joseph Alexander Brown
March 2020

CHAPTER 1

Introduction

Games provide a stage to be diverse. A player is an explorer embarked on an ambitious journey, a thief disguised as a detective, a sorcerer casting a spell, a saviour of the kingdom and a trader of trinket and treasures. Games with all of their miracles and magnificence is a player's passion.

This book celebrates player diversity. The first step toward celebrating diversity is *acknowledging* it. The influential capacity of games require a significant understanding of its players. What are player perceptions about design elements, what are their affordances, what do they like aesthetically, and such are the questions to be investigated.

All chapters in the book are dedicated to understanding players and their preferences for game design. Each chapter demonstrates a methodology to identify player preference trends and affordances for game objects. In short, the whole book is a process of unfolding mysteries of player interactions with games.

Lao Tzu states, *A journey of thousand miles start with one step.* The endeavors to understand the most miraculous creature, a human, is a journey, not comprising of a thousand miles, but a lot more and never ending. The most astonishing aspect of human beings is their non-static nature. However, some patterns have been constant since human behavior was first recorded and passed down to the present through tales, art, music, and poetry etc. One of these constant behaviors is the profound attraction to beauty. Beauty, an element of human fascination, has been endeavored to be created, been fought to possess, and aspired to retain. What is beautiful is a subjective question. The answer to this question has underlying reasoning, logic, perception, imagination, and desire. With all the diverse definitions of beauty, there is one non-conflicting opinion that beauty is *pleasant*.

Players are diverse and what gives them a *pleasant* playing experience has many aspects. To identify these aspects, this book presents empirical studies to determine player preferences, perceptions, and attitudes toward game objects and design. The studies are concentrated primarily on player perception and recent trends in player's likeness for new and unusual designs. For this book, we recommend a Statistics and Introductory Game Design course as a prerequisite for undergraduate upperclassmen and graduate students.

A brief introduction to all chapters' central theme is as follows.

Chapter 2: Affordance Theory and Game Design

This chapter discusses affordances from Gibson and Norman's point of view. The goal is to create designs that best suit user needs. For this purpose, the playtest methodology demonstrated in the chapter initiates and supports the process of understanding player perceptions and affordances.

Chapter 3: A Focused Conversational Model for Game Design and Playtest

Considering game objects as a target for learning process, this chapter details Context-Activity-Reflection-Documentation (CARD) model for playtesting. The Focused Conversational Model or Objective–Relational–Interpretive–Decisional (ORID) model supports designers in gathering player feedback and experience throughout the playtest. The chapter demonstrates CARD and ORID model with a playtest experiment.

Chapter 4: A Designer's Reflection on Game Design Considering Players' Emotions

This chapter demonstrates questions based on the iterative process. This process enables a game designer to reflect on game design to identify players' probable emotional outcomes at an early design stage.

Chapter 5: Age and Play

This chapter demonstrates a playtesting process according to the methodology explained in Chapter 2. The playtest has been conducted to identify player affordances for game design among two different age groups.

Chapter 6: Gender and Play

This chapter focuses on the gender of the player to investigate player perceptions of game design as well as their affordances for game objects.

Chapter 7: Player Perceptions of Odd-Shaped Dice for Dungeons & Dragons

This chapter discusses dice design and player perceptions of fairness and usability with classical and unusual designs.

Chapter 8: Dice Design and Player Preferences for Colors and Contrast

While dice being manufactured in a variety of colors and contrast, this chapter identifies recent trends in player preference of colors in dice.

1.1　HOW TO READ THE BOOK

The following Figure 1.1 describes the reading pattern.

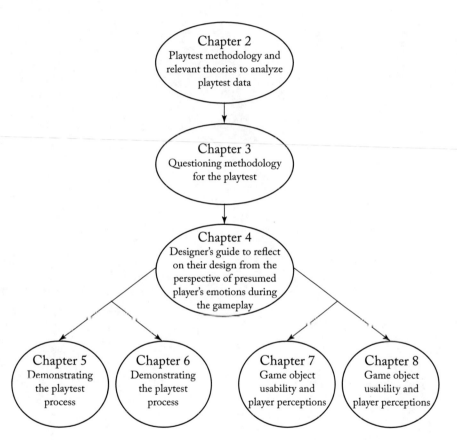

Figure 1.1: Reading sequence.

CHAPTER 2

Affordance Theory and Game Design

This book deals with players and their interactions with games. In consideration of human interactions, affordances are indispensable. Affordance is a word used to describe an interaction between people and the environment. The Theory of Affordances examines an individual's mental model and perception about the environment as well as interaction experiences.

This chapter entails two most popular concepts on affordances and a design testing methodology allowing for the investigation of user affordances for the design.

2.1 JAMES JEROME GIBSON AND DONALD NORMAN'S AFFORDANCE THEORIES

The psychologist James Jerome Gibson (1904–1979) introduced the word *affordance* (plural: affordances) to describe all potential interactions between the person and the environment. Gibson states:

> Affordances of the environment are what it offers the animal, what it provides or furnishes, either for good or ill [Gibson, 2014].

Gibson's notion of affordances points to all action possibilities concealed in the environment irrespective of the individual's ability to recognize them. As described,

> An important fact about the affordances of the environment is that they are in a sense objective, real, and physical, unlike values and meanings, which are often supposed to be subjective, phenomenal, and mental. But, actually, an affordance is neither an objective property nor a subjective property; or it is both if you like. An affordance cuts across the dichotomy of subjective-objective and helps us to understand its inadequacy. It is equally a fact of the environment and a fact of behavior. It is both physical and psychical, yet neither. An affordance points both ways, to the environment and to the observer [Gibson, 1986].

Gibson posits that affordances can have subjective as well as objective dimensions along with other aspects. Therefore, an affordance of an object or an environment might possibly differ from one individual to another. After understanding the concept of affordances, the next step is to discern how affordances study help in creating a valuable setting for an individual especially considering object interactions. Donald Norman has provided an extensive guidance in this dimension.

Donald Norman is involved in the field of usability engineering. In context, when Norman examine affordances, he emphasizes *Human-Centered Design*. Norman states,

> Human-centered design is an approach that puts human needs, capabilities, and behavior first, then designs to accommodate those needs, capabilities, and ways of behaving. Good design starts with an understanding of psychology and technology. Good design requires good communication, especially from machine to person, indicating what actions are possible, what is happening, and what is about to happen [Norman, 2013b].

Norman bases the principles of needs, capacity, and behavior on previous examinations of emotions [Norman, 2004]. Emotions have an impact on our reasons for purchasing habits. People's preference is affected by emotions they get from things they buy and in some cases they prioritize emotions over practicality of the object. This can create gaps between the design that attracts a buyer and that buyer's needs. Human-centered design is achieved by having the understanding of the humans for whom the design is intended to be used. Where *"understanding"* is a general word, it can be broken down into information about their capabilities and awareness of their interactions with a similar design. Norman has given a framework for understanding the process of human action. Forming a conceptual model, it consists of seven stages [Norman, 2013a]. These include in order:

1. *Goal*—forming the goal
2. *Plan*—the action
3. *Specify*—an action sequence
4. *Perform*—the action sequence
5. *Perceive*—the state of the world
6. *Interpret*—the perception and
7. *Compare*—the outcome with the goal.

These seven stages of action are dependent upon human factors such as an individual's age, experiences, gender, etc. As Norman highlights, most behaviors do not require going through all seven stages in sequence and there can also be numerous sequences iterated holistically.

The seven stages of actions are defining an individual's affordances with the environment. This leads to the core: we need to look into affordances that are the pure outcomes of the human factors (past experience and age, etc.). Elaborating on this topic, Norman is associating the seven stages of action to the three levels of processing (i.e., Visceral, Behavioral, and Reflective) within the brain and suggests that design must take place at all levels of processing. As explained by Norman,

> At the lowest level are the visceral levels of calmness or anxiety when approaching a task or evaluating the state of the world. Then, in the middle level, are the behavioral ones driven by expectations on the execution side—for example, hope and fear—and emotions driven by the confirmation of those expectations on the evaluation side—for example, relief or despair. At the highest level are the reflective emotions, ones that assess the results in terms of the presumed causal agents and the consequences, both immediate and long term. Here is where satisfaction and pride occur, or perhaps blame and anger [Norman, 2013a].

Norman suggests that reflection is the most important of the levels of processing to the designer because the emotions produced at this level are enduring.

Having informed Gibson and Norman's ideas of affordances mostly in their own words gives a sense upon affordances, these are all action possibilities from Gibson's notion. Norman emphasizes upon studying human factors so that the users are led to correct affordances for an intended design usability. A simple example would contribute to clarity.

Imagine you are an alien who has just come to visit Earth. An object has caught your attention—it is known as a pen to the Earthlings, such as in Figure 2.1. Can there be any guarantee that an alien who has never seen a pen before will correctly guess the intended usability of this object?

A pen is a simple example for this era as this is a common object for everyday use. Even this plain and universal object is giving illustrations of unintended design usability. The alien may see it as an object for fastening hair, a tool for poking and pointing, a human chew toy, or stirring coffee, etc. There are medical texts which examine a pen in emergency tracheotomies. Pens have been shown as improvised weapons in movies such as *Grosse Pointe Blank* and *The Borne Identity*.

A pen designer still has a great deal to work on to understand maximum possible affordances if they want to promote the usability of the product or inhibit unplanned interactions. A pen affords an action of hitting which is not intended by the designer. The addition of cosmetics to accentuate features of the pen requires a thorough analysis of each angle's affordances. A pen affording actions other than writing explains Gibson's notion as all action possibilities in the environment. In this example, Norman suggests investigating human factors and analyzing all possible interactions to limit the ones unplanned by the designer.

The subject of great significance is what proceeds after the designer has recognized possible interactions that they would like to limit or interactions they would prefer to happen in a certain way. The answer has been detailed by Norman in terms of signifier addition. In the field of design and usability, wherever you will come across the word *affordance* you probably will encounter the word *signifier* too. *Signifiers* are not the same as affordances. Norman distinguishes

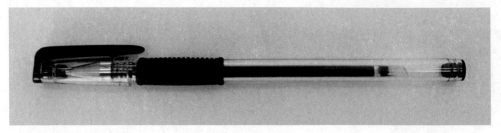

Figure 2.1: A conventional pen.

between signifiers and affordances as affordances determine what actions are possible and signifiers inform where the action should take place. Some affordances are perceivable, others are not.

Norman further describes that "Perceivable affordances often act as signifiers, but they can be ambiguous. Whereas signifiers signal things, what actions are possible and how they should be done. Signifiers must be perceivable otherwise they fail to function" [Norman, 2013b].

In the example of a pen, the nib informs the user which direction a pen should be held to perform the action of writing. If a pen has a cap, it gives the indication that a nib is inside the cap. In the case of a ballpoint pen, a clicker at the top is a signifier for drawing out a nib; a clean design and a divide at the top is a signifier of a rotating barrel to draw the nib.

Not all affordances require signifiers, as affordances can be perceivable too. Whether or not affordance is perceivable by the audience depends on the audience, so to be on the safe side, a designer must add a signifier if they intend to prohibit wrong usability and during the discovery process, prevent error. Norman refers to forcing functions to be added in the environment or the object, which is a means to promote the desired behavior by the user or to prevent incorrect actions. Forcing functions include *Interlocks*, which forces the function to take place in the correct pattern, *Lock-Ins*, such as preventing user from prematurely stopping an action, and *Lockouts*, referring to prevention mechanism from entrance into an unsafe space. These mechanisms in a design will lead a user toward the correct understanding of the conceptual model.

In order to have an effective design, Norman emphasizes conceptual models—a conceptual model is an explanation of how something works. Signifiers, affordances, constraints, etc. altogether provide a conceptual model of the system. The following quote is worth mentioning:

> The designer's conceptual model is the designer's conception of the product, occupying one vertex of the triangle. The product itself is no longer with the designer, so it is isolated as a second vertex, perhaps sitting on the user's kitchen counter. The system image is what can be perceived from the physical structure that has been built (including documentation, instructions, signifiers, and any information available from websites and help lines). The user's conceptual model comes from the system image, through interaction with the product, reading, searching for online information, and

from whatever manuals are provided. The designer expects the user's model to be identical to the design model, but because designers cannot communicate directly with users, the entire burden of communication is on the system image [Norman, 2013b].

The designer's job is well done if the system image created for the user is the same as a designer see themselves. To achieve this goal, users' perception about design elements because of their past usability experience or via elements of the design is indispensable to take into account.

Human-centered design and user affordances are closely related. Human-centered design considers human factors and cater to users' needs. Where human-centered design is a big domain, *intuitive* design is a subclass of human-centered design. An intuitive design comprehends that the design through its elements and features helps a user in exploring the usability pattern. The difference between human-centered and intuitive design is in what questions we are focusing on. If the goal is to give the user an ease in usability and decreasing frustration while interacting with the design then the design motive is intuitive. When the goal is also to consider the emotions and type of experience, etc., design domain becomes broader than intuitive design. In this book, the concentration is on achieving the process of human-centered design with an emphasis on intuitive design.

2.2 HOW CAN AFFORDANCES BE INVESTIGATED?

The discussion upon affordances and human-centered design gives the realization that a designer must take into consideration human factors while designing. This might seem obvious, but makes strong, yet subtle, demands for the physical design of objects, as well as the pragmatic design of how those objects are used. Design is therefore a social science, as much as it is an engineering problem. We believe in both: Gibson and Norman's notion of affordances for design enhancibility.

Norman frames the practical problem of design for humans, and his solution to closing the gap might best be found in how he updates the theory of affordances. Gibson states that affordances are part of the environment and are independent of an animal's perception. Affordances are always there to be perceived even if an animal does not recognize them. Norman alters the scope of Gibson's affordances by making affordances interactive. The concept of affordances can be used to evaluate any design.

Gibson and Norman's affordance concepts together provide a solution toward increasing design desirability. Gibson's notion cannot be ignored as it gives a huge (if not whole) set of potential design affordances (intended or intended by the designer) depending upon a user's design interaction and discoverability. Norman then guides toward comparing the gap between real (as intended by the designer) and perceived affordances and introducing signifiers and locking functions to decrease this gap. This put forward the big question: *How can Affordances be investigated?*

Depending upon the design category, the answer can be found through surveys, user testing and interviews, etc., with the target audience. Though these are helpful approaches, we propose a process that allows insight into user's affordances both from Gibson and Norman's point of view. The process involves letting a user interact with the object but *in the absence of any guidance books or instructions.* The following section describes the process with concentration on games, that enables us to investigate player affordances.

2.2.1   PLAYTEST

Game design exploration and investigation is an old practice. As an example, in the Persian *Book of Kings* (*Shāh-nāmeh*), the great poet Firdausi gives an interesting account of how chess made its way from India to Persia [Warner and Warner, 1909].

> As the story goes, in the 6th century, the raja of India sent the shah a chess set made of ivory and teak, telling him only that the game was "an emblem of the art of war," challenging the shah's wise men to figure out the moves of the individual pieces. Of course, to the credit of the Persians (this being a Persian story), one of them was able to complete this seemingly impossible assignment. The shah then bettered the raja by rapidly inventing the game of "nard" (a predecessor of backgammon), which he sent back to India with the same challenge. Despite its simplicity relative to chess, the intricacies of nard stumped the raja's men. This intellectual gambling proved to be extremely costly for the raja, who was obliged to pay a heavy toll: two thousand camels carrying "Gold, camphor, ambergris, and aloe-wood,/As well as raiment, silver, pearls, and gems,/With one year's tribute, and dispatched it all/From his court to the portal of the Shah [Yalom, 2004]."

Fifteen centuries later, Daviau used the same process by presenting a game to some participants without the rule book. The focus was on investigating the intuitiveness of game design for the people who had never seen that game before. Daviau states:

> Rules should not explain a game; they should only confirm what the rest of the game tells you [Daviau, 2011].

In this book, the methodology for affordance and design investigation has been adopted from Daviau's anecdotal study in which some games were given to people who had not seen those games before. They were asked to figure out the rules of the game without the help of rule books in a specified amount of time [Daviau, 2011]. However, Daviau did not elaborate on any findings from this study. This process when turned into an experiment opens the door to user's perspectives toward design elements and their perceptions.

Games are a complex object, especially board games, that come in a package, and when opened by the player, unveils another world where a player can immediately see the settings of the world without knowing the rules to live in it. The player sees this setting without prior knowledge if the rules or game mechanics will match the objects against their cultural expectations and perceptions about rules and goals. The playtest setting includes the following steps.

1. The experiment starts with a questionnaire. The participants fill in information such as their age, gameplay experience, how many hours (approximately) they spend playing games in a month, and which kind of games they like to play.

2. The next step is presenting the board game without rule books. If the packaging has rules written on it, it is also removed and only game objects are presented to participants.

3. In a specified amount of time (usually thirty minutes), participants analyze the game objects and fill in further questions regarding the game.

4. The questions to be answered after analyzing the game elements include:

   - How many players can play this game and who goes first?
   - What are the play mechanics/rules?
   - What is the winning condition?

The participants must not have seen the game under test before. This is crucial to the experiment as the designer needs to know the player's affordances and any prior knowledge about the game rules will not give beneficial information to the designer to improve the design. After the playtest, the observer asks participants questions about their interaction with the objects and design. The questions are grouped into four categories: Objective, Relational, Interpretive, and Decisional (ORID). The questioning process is detailed in Chapter 3. The playtest as a whole is a CARD (Context-Activity-Reflection-Documentation) model such as setting up the game and questionnaire as well as instructions for the playtest are the context part. The player then proceeds with the activity, such as playtesting the game. The observer conducts reflection as an ORID model and finally the observer takes away the results of the playtest and the interview phase as their documentation.

The experiment may apply to any product or environment setting. This methodology has been successfully applied on board games. Chapters 5 and 6 in this book demonstrate the playtest process and analysis based on participants' age (Chapter 5) and gender (Chapter 6). The goal is to make rule books as thin as possible.

Less reliability on rule books and game design research is critical as game application areas are increasing. Games have proven to be powerful tools of simulation of war in historical settings. Chess is perhaps one of the most well-known wargames of historical use, with simulations of the Indian armies in the 6th century. However, it would be hard to find a 19th century Prussian officer which had not been introduced to Kregsspiel (literally "war game" in German). In the Second World War, wargames were used by the British Navy for the modeling of naval convoys in the Atlantic and examining potential U-boat tactics. Today, combat simulations of all forces are not only used by the military, but the technologies have found commercial success as simulators for emergency forces, aircraft simulations, and in entertainment. The role of games as entertainment has been a secondary use of the technologies, and the idea that "games are only for children" has no historical relevance, and is incorrect in present day based on demographics of game players, both digital and analog.

Research trends are advocating design that does not need much guidance from rule books. Browne [2015] discusses game design patterns and principles that can lead to games which are enjoyed by the players. The game objects should speak for themselves, which means to embed the rules in the game design itself. This encourages design elegance and clarity and hence gives the player an enjoyable experience. Some examples from games have also been demonstrated by Browne, where the game design has a "poke-yoke" effect—poke-yoke refers to reducing the player error by making a design that itself inhibit such errors. Yermolaieva and Brown showed that even with objects as simple as dice the design could influence play in a game due to errors caused by the incorrect reading [Yermolaieva and Brown, 2017]. The game design investigation method with the removal of rule books has been demonstrated by Brown et al. [2019]. It is observed that features such as game mechanics, game colors, objects, etc. contribute to developing a player's conceptual model about the gameplay. Moreover, Aslam et al. [2018] has applied the same playtest methodology (such as rule books removed), for examining player affordances and game intuitiveness for different age groups. The removal of rule books gave insight into a player's affordances and perceptions about design elements. The research analysis such as identification of common interaction patterns of players within the same age group is advantageous for game designers to initiate an intuitive design process considering the player's age and preferences.

Our work details affordance investigation process. However, the concept of affordance within the game design sphere is not new. Linderoth [2013] describes gameplay as a process of perceiving and choosing affordances. Mateas [2001] talks about material and formal affordances in the gameplay. Furthermore, works such as Young and Cardona-Rivera [2011], Vera and Simon [1993], and Suchman [1993] are an informative read on application areas and points of view on affordances.

2.3 PLAYTEST LEADING TO GIBSON AND NORMAN'S AFFORDANCES ANALYSIS

We endeavor to introduce a research methodology that can give the designer an understanding of player's affordances with the game design and objects. The goal is to have knowledge regarding what interactions can be carried out with the game design (which is Gibson's notion) and what are the interactions a designer intends to generate as well as what interactions players perceive to be existing (Norman's notion).

Ultimately, this experiment is meant to lend insight for video game design. A board game is used because board games provide an abstraction of video game elements. Board games differ from video games in two useful ways for a study about games and rules.

Players can start their analysis at the end of the game, or from the middle, and work their way backward. Every piece that they could interact with is available to test in any order they wish. Players can rearrange the board game's pieces and the narrative of the game is not necessarily obvious.

On the other hand, a video game only presents whatever can fit on a digital screen at one time. Video game narratives are tightly controlled and cannot be rearranged or played in a different order than the game intends. Video game components are often hidden and their discovery is part of the gameplay.

Board games are, for this experiment, games that contain no electronic components. A video game controller makes an experiment like this one infinitely more complex. The number of buttons and movement options are intractable. Does a pause button and access to menus count as part of a game or separate because the game is paused? How long would it take a player to discover the Konami Code by systemically pressing buttons to determine the actual rules of a game?

Board games are different is that they come in a package that can be opened. Once opened, the board game is an exploded model of itself. Players access the game in a way that should make any engineer or mechanic jealous. It is possible to observe players make inferences about the pieces and their purpose, and test the pieces against their perceptions and expectations about rules and goals. Video games are, in this way, not interactive because their pieces are hidden and their controls are so linear as to be a form of deus ex machina; they are not discovered, through the curiosity of the player because so few pieces of the game's narrative are visible at a time. While part of the pleasure of a video game is the unveiling of strategy as narratives develop, the process of playing a video game does not reveal the critical thinking of a player the way unboxing a board game does.

The methodology of this experiment is based on an analogy that unboxing a board game is like seeing all the components of a video game's narrative at once. This abstract ideal of narratives and playing pieces is one of the features that distinguishes games of all types from literature, video, sculpture, or painting. However, the ability to unbox and test pieces in different combinations is a lot like the ability of poets and painters to rearrange the components of narratives to create new understandings and test new narratives. In this way, board games, and thus games in general, are similarly creative endeavors.

Considering affordances, the methodology proposed here enable game developers to identify user's perceptions of game objects to make a comparison between real affordances (as intended by the designer) and perceived affordances of these objects. Norman defines perceived affordances as how an object may be interacted with based on perception [Norman, 2013b]. The playtesting methodology, involving the removal of rules, provides the absolute conditions for a player to make actions according to their perceptions. A comparison between the player's perceived affordances and real affordances of game objects allows the analysis of game object designs to see which objects have led the player toward the correct mapping of affordances and which objects persuaded the player toward an incorrect or opposite mapping of game mechanics—an anti-pattern.

The playtesting environment discussed here is not that of actual in-competition gameplay. An individual is examining the game as if they first opened up the box and initially attempts to understand what is in the box. This allows for greater freedom in the player's mind as other players are not enforcing a set of known rules. Gibson defines affordances to be all action possibilities latent in the environment irrespective of an individual's ability to recognize them [Gibson, 2014]. In a non-competitive environment, playtesters are naturally propelled to try out various actions with objects that could make them recognize affordances which were not observable by their perception initially.

A game design from Gibson's point of view of affordances is significant to understanding all possible actions an object offers. At the same time, the removal of rule book has brought a change to the Game's environment. This helps to measure the change of object based on the existence and non-existence of rules. Moreover, the study of game design from Norman's point of view of affordances is also crucial to understand how well the game design supports its play mechanics.

Our adopted methodology for playtesting investigates game designs from both Gibson's and Norman's approach toward affordances. Knowledge about players' perceived affordances helps to measure the impact of human factors on players' understanding about game actions that can affect their level of enjoyment with the game. Hence, games designers can account for the extent, as well as the impact of human factors to achieve the intended purposes of the game. Moreover, as players are propelled to try out atypical actions in a non-competitive testing environment with the absence of a guidance manual or a rule book, this methodology also gives knowledge about all possible affordances with objects and allows for the improvisation of new games designs.

CHAPTER 3

A Focused Conversational Model for Game Design and Playtest

The focused conversation model was first developed by United States Army Chaplain and Art Professor Joseph Mathews as a method of art appreciation and reflection [Stanfield, 2000]. The conversational approach eschews the idea of a single expert in the teaching model and instead relies on the assumption that there is no universal truth to the appreciative process. The only method to come to meaning is to have a series of viewpoints. A truth is based on observing several subjective opinions, and not a result due to the existence of a truth objectively. When looking at the world of games—avoiding the common folly of stating analog or digital games are completely unrelated design tasks—we realize the need for the development of such shared truths in the rules sets. That being a set of rules by which the players will abide by setting out the expectations for play and those actions which are out of bounds.

Suits defines a game as "the voluntary attempt to overcome unnecessary obstacles" and states further that "to play a game is to attempt to achieve a specific state of affairs [prelusory goal], using only means permitted by rules [lusory means], where the rules prohibit use of more efficient in favor of less efficient means [constitutive rules], and where the rules are accepted just because they make possible such activity [lusory attitude]" [Suits, 1978]. This definition is an operationalization of games as rules sets. However, the play does not always have explicit rules.

Children at play also show these implicit structures [Castle, 1998, Golomb and Kuersten, 1996]. They will start to form the basis of the space about them and the objects available to them. They quickly define a basis for the collective play, toss aside previous methods, and will declare other actions outside of a norm as "cheating" or "unfair" in order to maintain a collective law. The children in their attitudes set forward a lusory means as a matter of play and develop manners to ensure constructive rules are not unduly influencing the play session to another player's advantage.

Such methods have also been used implicitly by game designers and developers, such behaviors as those engaged in the Blackmoor campaign [Peterson, 2012] which would serve as the foundation for Arneson and Gygax in the codified rules of Dungeons & and Dragons (D&D). When the campaign was ongoing, Arneson was asked to provide the rules for the campaign only to hear the response of "Rules? What rules!?"[1] Of course, there were rules; they were implicitly known by the players, defined over a multitude of play sessions, drafted, agreed upon, ratified, and revised. When an event occurred which was outside of the agreed-upon limits of the game, it would be disputed by the community and a new rule would be developed. Arneson would even eventually go back on the statement and present a set of rules to Gygax, and other elements such as the use of the 7-die polyhedral set instituted by Dave Wesley would also be adopted from Blackmoor [Tresca, 2011a]. However, the Blackmoor system was not "constructed with any consideration to instructing someone who had not already experienced it," and thus Gygax would mould Arneson's "20 or so pages of handwritten rules" through a process of "adopting, expanding and repurposing them" [Peterson, 2012]. This was not a new creation from the request, but a codification of a collective oral history of a community's laws.

In the development of a game, there is a balance. Rules create enjoyable games where players have choices of actions, their fairness to the players, and players find a clear direction to make a reliable strategy. However, a game with too many rules limits interactions; it becomes too hard to understand what is a legal interaction, and the rule become self-contradictory and internally consistent. Further, games have narrative and themes outside of the mechanical actions of the rule. Rules should reflect these narrative choices and themes of the game.

The focused conversational model allows for an examination and reflective process, and it works effectively when combined with an activity-based learning model. The CARD model is an activity-based reflective process of lesson planning. It has been used as part of the set of training processed which is used in the Instructional Skills Workshop (ISW). The ISW is an internationally recognized/utilized best practice for faculty development for instructional skills and lesson planning [Dawson et al., 2014, Macpherson, 2012, Zhirosh et al., 2019].

Section 3.1 reviews the lesson planning model of CARD with an examination of its applicability to playtesting. Section 3.2 overviews the focused conversation model or ORID. A case study of a game rule process is examined in Section 3.3 with a sample set of facilitator questions. Section 3.4 examines the findings of the case study. Finally, Section 3.5 gives conclusions and a future direction for games educational methods.

3.1   THE CARD MODEL

The Context-Activity-Reflection-Documentation (CARD) model of planning focuses on providing an experiential and outcomes-based learning environment. The goal of the model is to reflect a shared activity, which is modeled for an experiential learning process, especially for those with an outcome-based learning model. In terms of game design and design of rules, there is no clear, measurable objective, yet there is an expected outcome of finding flaws in the design or understanding of players using features of the game's mechanics.

---

[1]Attributed to *Different Worlds* issue 3 in July of 1979 by Peterson's account.

3.1.1   CONTEXT

The contextual phase readies the space for learning and sets the plan for the activity to come. It sets some expectations as to the method but does not intend to limit the scope of the participants to provide feedback. The context should make the activity clear for all participants in terms of their roles, the method of which they will use to engage with the artifacts of the game objects, and what rules will be introduced.

3.1.2   ACTIVITY

The participants engage in the outlined activity. The facilitator is monitoring and making observations of the behaviors in order to have an objective record of the event. The activity may also be filmed or otherwise recorded: keystroke logs, actions made in the game, etc.

3.1.3   REFLECTION

The participants are asked to reflect upon the previous activity. This phase is primarily where the discovery or learning event occurs. The role of the facilitator is to elicit ideas from the participants. This can be done in several manners. The facilitator can use a survey-based approach giving a series of closed questions (with generally quantitative results) and open questions (to give qualitative results). This may also adopt the format for a larger group as a conversational model, such as the ORID Model.

3.1.4   DOCUMENTATION

The end phase is a collection of the documentation for both the facilitators and the participants. In a learning environment, this is valuable to students as an artifact of the process to be reviewed for tests and projects or as an example for their future processes.

   As a design tool, the documentation is for the designer to have a record which allows for both analysis for the developer for improvements. Moreover, it allows the participants an appreciation.

3.1.5   PLAYTESTS

To contextualize this model into the application in playtesting a new game, we need to see the game objects itself as the target for a learning process. The expected outcome of our playtesters is an understanding of the game objects, rules, interactions, and strategies. For those who are just seeing the game from the first time, we are interested in the learning process of a novice in understanding possible conceptual models of gameplay. For those who have interacted with previous editions of the game, the designer is interested in seeing the evolution of thinking about more complex conceptual models and moving from short-term tactics of playing into more long term strategies.

3.2   THE FOCUSED CONVERSATION MODEL – ORID

The focused conversation model or the Objective–Relational–Interpretive–Decisional (ORID) model is applied in the following manner. It first examines the data—Objective. Then makes a call for personal feelings and reactions—Relational. It encourages the group to delve deeper—Interpretive. Only then does it require the group to make either a collective or personal statement as to what the situation or object has taught them or what actions should be made next—Decisional. To follow is a more detailed look at the model and some example/sample questions used in our examination.

3.2.1   OBJECTIVE

Objective-based questions are those which explore the situation or object based upon the objective facts without recourse to emotions, beliefs, and interpretations. The goal of this stage is to have a collective and clear agreement as to the question of "what happened?". It is often the case in this stage that facilitators need to keep participants clearly on task by limiting the comments at this stage. It is a natural want of the participants to want to skip over this step in their responses and move on to their beliefs and interpretations of their actions. This is folly. Allowing this will lead to confusion for other participants and will lead to statements about actions without everyone understanding what happened from the various perspectives and will not allow an objective view of the situation or object to emerge.

3.2.2   RELATIONAL

Relational questions examine the feelings and base-level thoughts of the participants. The role of this level of questioning is to have the immediate and quite often personal reaction to the data. This is often the emotional response suppressed in the objectives level of the questioning. This level examined the surface relations between the facts.

3.2.3   INTERPRETIVE

The *interpretive* level of questioning examines the meaning of connections. These questions examine the values of the participants and the implication of the thoughts about the situation or object. This stage of questioning has been built into by the previous questioning stages allowing for a firm foundation for insights to emerge.

3.2.4   DECISIONAL

The *decisional* level examines, based on the interpretations, what actions should now be undertaken. The questions of this phase create resolutions and close the conversation while laying out the next steps. This stage also naturally lends itself to the documentation of an action plan.

Figure 3.1: Instructor gives the *Context* of the process. Photo by Lesia Poliakova, used with permission.

3.3   CASE STUDY

3.3.1   CONTEXT

The players are told to join in on the development of a game. The room is cleared of any obstructions and facilitator(s) lead the group(s) of 5–10 players arranged into circles; see Figure 3.1. They hand several game objects to the players. The players are then informed that they will be constructing a game rule by rule, each player in the circle receiving a chance to add a rule. If at any time the group believes that the rules have become too complex, the game is ended, and the number of rules is recorded. The game is then reset to having no rules.

3.3.2   ACTIVITY

During the activity stage, the groups are arranged and the facilitator provides a set of game objects, such as balls, mats, etc. During the gameplay, the facilitator is observing the process and making mental and perhaps physical notes on the actions made in the game. Their note taking aim is to allow for better questioning rounds during the reflective conversation. In addition,

Figure 3.2: Students engage with the development of a rule set adding one rule at a time during the *Activity*. Photo by Lesia Poliakova, used with permission.

the facilitator is there to ensure participant safety and to be able to answer any questions about the process from participants. They also act as a time keeper for the rounds of the playtesting between rules changes and for the activity overall.

3.3.3   REFLECTION

The reflection method has two parts. The first is at the end of every round of the game, see Figure 3.3, in which the players declare the number of rounds to be over as the game has become to complex. The facilitator will check with the group in order to note anything they would like to have a remembrance upon and examines the comfort of the group before the next round.

In the grand reflection phase, the ORID model is used with each participant keeping their own record and the facilitator writing a group finding on a whiteboard. Each individual is asked the following questions as a minimal set—with deeper probes by the facilitator.

Figure 3.3: The group meets again to *Reflect* on the rules sets created. Photo by Lesia Poliakova, used with permission.

Objective Questions
- What instructions were given by the facilitator?

- What objects did you use?

- How many rules did you make?

- What were the rules you produced?

- What rule caused you to end the game?

Relational Questions
- How did you feel when engaging in the activity?

- Why did you choose these objects?

- As the number of rules increased, what did you notice?

- What were your feelings about the rules you produced?

- How did you feel as the game went on toward the end?

Interpretive Questions
- What rules were better for the game?

- What rules were contradictory and how did you solve this problem?

- Why did that rule cause the game to end?

Decisional Questions
- What rules would you want to keep in future games?

- What rules would you want to avoid in future games?

- What did you learn about rule systems?

3.3.4   DOCUMENTATION

Each participant has their notes and record of the playtest questions, and the grand collection is sent as a picture of the whiteboard to the class.

3.4   FINDINGS

3.4.1   LATER ROUNDS COULD GO LONGER

In the multiple times this process has been run the initial rounds of the game creation lead quickly to games where the participants wish them to stop. Later examples of games will have more rules added before called to end.

The speculated reasoning is that as players learn the actions which lead to a reduction of the fun, fairness or are contradictory. It was noted by several participants that with a few rules the game was not fun due to the lack of challenge and too many rules lead to frustration. For example, there are physical limitations of being to stand on one foot, while dropping the ball, shouting the name of who passed the ball, and then passing the ball to someone who had not yet passed the ball. There is also the problem of the cognitive load of knowing when and in what order to pass the ball.

3.4.2   DEVELOPMENT OF ANTI-RULES

In multiple instances, players would engage in the creation of rules which would counter what they perceived as unfairness or cheating in the game. Having a short period of play between the addition of rules allowed for a clear understanding of the effect of playing the current ruleset.

In one illustrative instance, a rule was instituted which eliminated players tossing the ball if the ball was not caught. This quickly leads players to intentionally drop the ball or refuse to catch the ball, wanting other players to. After a few such eliminations, a player with the ball just held it, much to the dismay of those assembled who realized that he was attempting to avoid elimination by refusing to play.

This was then countered in the next rulemaking round by a rule which required a player to pass the ball under a five-second shot clock. That, while fixing the issue of the player just holding onto the ball, reverted the game to the previous state of having an elimination on each pass of the ball, still considered to be unfair. The next rule added was to counter this deficiency by eliminating both the passing player and the receiving player when the ball was dropped. This process of creating rules and exceptions left behind the shot timer which would persist in the rules for the remainder of the game.

The vestigial nature of this rule would later emerge during the focused conversation. During the relational phase, it emerged to this group via the question of if this stage of the game rules was very frustrating, hard, and trying. The interpretive phase of questions examined these statements further, and it was decided that the shot clock was added to remove the immediate issue of the lack of action in the game, however, the longer-term fix of eliminating both the passer and receiver was what led back to the fun/challenge in the game.

3.4.3   FACILITATOR AS AN APPEAL

As a facilitator, there may be questions when these perceived offenses occur—that is, the facilitator as a court of appeal or a rule meta-maker. The practice adopted is to remind the participants that (1) they can collectively reset the rules any time to nothing with a consensus of the players, and (2) at the end of the play round if an exploitation in the rules is perceived, then the rule can be changed by the next rule maker unilaterally.

However, it is best if the facilitator engaged when rules declared are sexist, racist, exclusionary, or directed at individual players and not behaviors in the game. Thankfully, in the author's experience this has only occurred once in the formation of an anti-rule to prevent perceived cheating actions by a specific player, and a reminder of "well what would happen if someone else made the same action?" from a facilitator was enough to focus the team and revise the proposed new rule with no offense taken.

3.5   CONCLUSIONS

The process of the focused conversational model allows for a lesson which can examine a topic without clear objective-based outcomes. The model is inherently experience based, allowing for active learning. It also a reflective process which has been demonstrated to improve learning outcomes. The model also allows for a method, when used in conjunction with the CARD framework to be used in lesson planning, making the process of active examination of grams following a clear pedagogical process and allows for the replication in course syllabi, which games education is sadly missing.

The future development of education in games development is not in retesting the theoretical basis of the pedagogical models. There are existing good methods such as focused conversation and CARD which are well framed, used in several practices in the related field of artistic appreciation, and trained in accepted international pedagogical best practices used in many countries.

The future work should accomplish the following. (1) Put into practice clear and existing leading pedagogical models, first via the development of laboratory, industry, and research leaders trained in better distribution of their works to students. This interfacing with the training of upcoming graduate students and teaching assistant roles who have more contact hours with students in smaller group lab and tutorial settings were such techniques as the CARD and focused conversational models applied in their most useful context. (2) Address the development of exemplary classes using the models to fit with a games development curriculum. The need for a curriculum is perhaps the great challenge, a clear games development curriculum, as there is currently no generally accepted elements which separate a games design or development degree as a type of specialization. This conversation should be undertaken by practitioners and stakeholders to create a standard for the specialization and will need to emerge to move game design and development from what is now, an early academic field, into maturity.

CHAPTER 4

A Designer's Reflection on Game Design Considering Players' Emotions

Emotions and perceptions are associated with each play experience. Therefore, players potential emotional outcomes must be regarded at an early stage of game design.

The goal of this chapter is to demonstrate to readers how to design a game, taking into account player's emotional outcomes along with the designer's intended theme and mechanics. Playtesting is a crucial part of development process, however, a designer can obtain understanding about player's emotional journey at a point when they have only decided some broader details and ideas about the game. The player's mental process cannot be assumed without a playtest and investigation as players are diverse and so are their gameplay experiences. This chapter aims at helping designers to formulate the game template with possible emotional outcomes that they deem closer to players' mental process.

Game design starts with certain expectations regarding play experience. The designers' major goals regarding play experience can be player engagement, fun, excitement, enjoyment, etc. These elements are assessed when the game is in the playtest phase. The designers set mechanics, paying attention to the desired goals for the player experience. However, the mechanics themselves are too interconnected and maintaining a balance between game mechanics and designer's goals regarding player's emotional experience can be difficult.

Games are usually designed with a theme or mechanics setup as a starting point. However, only after the theme and mechanics are laid out, the player experience can be investigated. This chapter entails a process that prioritizes and ensures that gameplay can (to a high extent) achieve designer's goal for player experience and emotions along with desired mechanics and settings. This design process includes several questions for the designer. These questions help game designers reflect on the mechanics to identify players' potential emotional outcomes.

For the process demonstration, the assumption is that a game designer is starting the design process for a new game. The game mechanics are being investigated, considering emotions that the player is expected to experience during the particular stage of the game. In our demonstration, the questions of this iterative process are answered as the first person to support the understandability as if reader is part of the design process. Questions posed in the iterative process are answered assuming that no major work regarding game settings has been done and a hypothetical designer starts from basic ideas in their mind.

As an example, we suppose that the designer plans to have a classical storyline in which players are hunting for a treasure and the treasure is guarded by the Beast. The iterative game design process starts with the designer answering the first two questions about emotions and psychological state, they intent for players to achieve or not experience (dominantly) throughout and at the end of the gameplay. The remaining questions are related to a broader detail of game mechanics that once answered will be matched against corresponding designer intended player's emotions. The process goes as follows.

1. Game designer answers seven questions according to the current setup or ideas about the gameplay.

2. Once all questions are answered, the designer focuses on questions from 2–5 and analyzes possible emotional states a player can experience. The emotional outcomes (as inferred by the designer) are then matched with the answers of first and second questions.

3. The positives obtained from questions 2–5 (such as all intended emotional results that the designer speculates from the current settings) and negatives (such as emotional results, the designer speculates, that are not intended by them) are written down. The designer further progresses to create mechanics and settings while considering balance or some modifications in mechanics that can reduce negatives such as making them less dominant as the play proceeds.

Thus, the process reiterates several times and each time with progress in game design, until the designer is convinced that the developed game mechanics have coherence with the emotional outcomes, which are expected from gameplay.

4.1   QUESTION-BASED ITERATIVE PROCESS

The following hypothetical example provides the template for designing a game keeping player emotions and expectations as a design priority.

1. Which emotions do I want to dominate during gameplay?

   Answer: Considering this question, the hypothetical designer decides and writes down three major emotions or psychological states, such as:

   I want my game to evoke the following emotions during play experience:

   (a) Excitement

   (b) Curiosity

   (c) Engagement

2. Which emotional states do I want to be less dominating during gameplay?

   Answer: The three major emotions or psychological states that I do not consider desirable to be dominant in player experience are:

   (a) Disappointment

   (b) Fear

   (c) Helplessness

3. Does my storyline represent a fantasy or is a simulation of a real-world scenario? Which objects will be part of game design?

   For this question, the designer has to think of the scenario constituting gameplay environment.

   Answer: The hypothetical designer answers the question based upon current thoughts, such as:

   I want a fantasy setting; I want my players to embark on a journey toward an unknown valley. So far the valley only exits in books which were written 300 years ago.

   With regard to the main game pieces, the idea is as follows:

   I would include token pieces. Some of the token pieces will give some advantage to the player toward killing the Beast and some pieces will allow the player to introduce some strategy toward killing the Beast.

4. Do I want cooperation or competitiveness in the game?

   Answer: I want a competitive game with some ability to cooperate that provides benefit to both players in cooperation. According to the current ideas, players can fight together in the forest and then compete in the end to kill the Beast.

5. Which element must be stronger for my game: luck, strategy, or skill?

   Answer: I want luck to dominate strategy.

6. What is the winning point in my game?

   Answer: As a final goal, I want the player to discover the valley and get the treasure which has been guarded by a Beast for 300 years. The treasure is hidden in a cave and the Beast sits outside the cave.

7. What is the biggest adventure or threat in the Game?

   For this question, the designer writes down the general idea about the highlight of the game such as the biggest adventure they want the players to experience and which emotion the designer intends for this setting. The hypothetical designer answers.

Answer: I want player(s) to interact with the Beast and if they can not kill it; the Beast can kill the player based upon two conditions (to be decided later).

The second part of the answer about dominant emotion or psychological state for this setting is *engagement*. I want my players to feel strong engagement and excitement at this point.

4.2   REFLECTION ON SEVEN QUESTIONS PROCESS

The designer has written down broader aspects of the game design and there are still several elements and concrete details missing. This is sufficient at this stage as the idea is to have the designer obtain understanding of their expectations about the player's emotional experience.

Now the iteration begins and the seven questions process is revisited to cover all aspects of designer's goal for players' emotional experience. This allows the designer to reflect on their broader mechanics of the play and identify which emotions can result from these mechanics and how close these emotions are from the designer's goals about players' emotional journey.

4.2.1   REVISITING THE FIRST QUESTION

Getting the Emotions We Desire to Get

The first question sets the designer's goal of three dominant emotional experiences throughout the game. Going through the current answer set, the following is apparent.

1. "Excitement" element is expected to be attained from the storyline, such as players moving through forests, fighting strange species, and having unexpected encounters. Furthermore, the main adventure of the game that will lead to the winning condition (encountering the Beast) has the capacity to generate excitement and engagement states.

2. "Curiosity" can be attained through the Beast dealing aspect where a player knows the Beast's weakness and tries to work with either luck or strategy to defeat and kill the Beast.

3. "Strong Engagement" can be attained through the Beast encounter phase during which the player does not know the outcomes. Moreover, the player is naturally propelled to immerse in the gameplay with full concentration and engagement when they deal with the Beast through strategy tokens.

4.2.2   REVISITING THE SECOND QUESTION

Not Getting What We Do Not Want to Get

Upon analyzing Section 4.2.1, identify whether the results are undesired from the same settings. The hypothetical designer does not want the following.

1. Disappointment: Are we getting disappointment through any of the settings suggested in Section 4.2.1? The three elements in story settings to achieve excitement, curiosity, and

stress have the capacity to lead to disappointment. A player can feel disappointment when they do not succeed to conquer forest trials. They can feel disappointment when they are trying to find the luck token and they get the strategy token. Losing the game can deliver disappointing emotions.

The mechanics needed to balance the opportunities in order to avoid *disappointment* becoming overwhelming. This can be done by adding mechanics that give some probability to the player to get back to the game with the excitement of winning. Strategy token will help in this context. It is the matter of the playtest to identify if the disappointment led to motivating the player to play another round of the game or it led them to quit playing. However, this point must be addressed when the details of these play settings are thoroughly defined.

2. Fear: Upon analyzing the same settings again, fear is expected to come from all three settings. When moving through the forest, player and Beast interaction and unknown outcomes can all evoke fear. These settings require re-evaluation to decrease the possibility of fear dominance and increase the possibility of the desired dominant emotional state goals.

3. Helplessness: Curiosity and engagement enforcement settings can lead to dominant *helplessness* feelings such as a player might feel while figuring out the Beast's weakness during the Beast encounter phase. These settings needs to be revisited to decrease the probability of helplessness by adding player supporting strategies or luck elements in the game.

4.2.3   REVISITING THE THIRD QUESTION

The story line supports fantasy mode. The story has several aspects and corresponding game objects defined so far that were discussed in Sections 4.2.1 and 4.2.2 for emotional outcome assessment according to the answers given by the designer.

4.2.4   REVISITING FOURTH QUESTION

Based upon the currently decided settings, the game has a competitive mode where players are trying to collect enough points to get to the Beast and kill it. Some elements of cooperation are embedded in the scenario of players fighting with dangers in the forest and sometimes defeating a danger cooperatively (either in pairs or any number) gives points to each player involved in the fight. There exists no conflict with designer goals for desired player emotions in the current settings. The competitive element leads to undesirable emotions (disappointment, fear, helplessness) as discussed before. Mechanics can be modified to make these emotions less probable to dominate.

4.2.5   REVISITING FIFTH QUESTION

According to preferences, luck should dominate the strategy. Investigating settings in the current process, the luck is introduced through tokens that the player will randomly draw (conditions are not decided yet).

Going back to the first and second questions, the luck and strategy mechanics will influence the emotions mentioned in both questions. This question will be revisited when the conditions are written for applying the strategy. The luck (so far) is being considered by a random token draw and this can lead to disappointment, helplessness, and fear. Once the mechanics are fully written and in another iteration, this section is examined to balance between luck and strategy mechanics; thus, undesired emotions dominate the player's experience.

4.2.6   REVISITING SIXTH QUESTION

Does the goal of the game conflict with or support the designer's goal for the first and second questions? The answer at this first iteration is not clear as the whole mechanics are not written.

4.2.7   REVISITING SEVENTH QUESTION

The desired main adventurous element in the game, such as player and Beast interaction, support the first question (achieving desired emotional outcomes) and is consequently giving excitement, curiosity, and engagement (as this is also deciding player win/lose). Furthermore, it also invokes the second question's states where not achieving the desired results can cause *disappointment*. However, disappointment must not dominate during the play process and at the end of the game. Even after losing the player must have a positive experience that led them to be excited about the second round of the game.

Furthermore, investigating this question for the emotions, we do not want *fear* and *helplessness* to dominate. Fear exists as encountering the Beast is fearful and helplessness also arises from the introduction of the luck element. The player is *helpless* (partially or fully?) in this situation of randomly picking a card that can decide whether or not they are lucky.

4.2.8   FURTHER ITERATIONS

With the development of mechanics and story settings clarity, this process is reiterated to analyze emotional outcomes from the latest settings.

As for the question, how many times the process should be iterated? the answer is, as many times as the designer considers beneficial. As the mechanics develop, the analysis becomes complex. The first round of iteration gives answers to questions such as: Which emotions are assumed to arise from game settings?

Further iterations with more developed mechanics reveal and inform: How can undesirable emotions be suppressed by mechanics and/or by increasing the probability of desired emotional outcomes?

The process continues until the designer is convinced that they have analyzed the emotional aspects and outcomes from all design details including game settings, mechanics, and play progress.

4.3   INCORPORATION TO OTHER GAME DESIGN APPROACHES

Game design can start from mechanics development or theme setting. There is no one rule to design and it depends on the designer's ideas and preferred process. The seven questions-based iterative process presented in this chapter can be incorporated into any game design framework. It does not matter if theme to mechanic or mechanic to theme approach is being used or any other design methodology. The seven questions process gives the designer a view of the players' emotional outcomes with each game setting. As an example, the three act-based framework presented by Tidball [2011] states:

> Gameplay's first act—its beginning—is when the stage is set for conflict among the players. Battle lines are drawn and the players understand the dimension of the conflict.
>
> Gameplay's second act—its middle—is the meat of the struggle for victory. Each player constantly strives to establish a compelling and enduring edge over the others, so he can make a final push for victory.
>
> Gameplay's third act—its end—is the push for victory. One player or several players in succession, either and try and fail or try and ultimately succeed in sealing the deal and ending the game with their own victory.

Considering this framework, game designers are able to apply the seven questions-based process in all three acts. The designer can investigate emotional outcomes from three divisions: gameplay's first act, second act, and third act.

The designer without the presence of playtesters can speculate emotional outcomes that can be later confirmed or assessed via the playtests.

4.4   DISCUSSION

The questions-based iterative process assists game designers to prioritize and assess game design based upon their goals for player emotions. The designer reevaluates game settings and mechanics to assess the probability of getting desired emotional outcomes from the game players. In order to concretely estimate players' emotional outcomes, psychological concept awareness is required which might not be the expertise of game designers. Such efforts are recommended as they will lead to productive outcomes, however, not having enough resources to carry out such actions should not stop the game designer to reassess the game design considering expected player emotions as important as game mechanics that will set the ground for these emotions.

Players are diverse and will experience different emotions in the same game settings. Usability and interaction experiences are influenced by perceptions as indicated by Norman [2004]. Knowing this, an exact conclusion cannot be drawn for the whole game audience. However, this can enable game designers to estimate probable emotional outcomes from the player side.

The next step is to discern how close our assumptions are to the actual outcomes. This goes back to Chapter 2 where playtest informs about players' perceptions and emotional experience.

CHAPTER 5

Age and Play

Age is a gift that must be embraced in its entirety. Age is not just a number tagged to an individual, calculated from the year of birth as it holds in itself experiences, memories, accomplishments, hopes, wishes, aims, and millions of treasures given by life. It is a beautiful truth and a mystery as well. It is a simple notion yet so complicated. No one has ever been able to confine age in one definition. For some, it is the memories gathered and for others it is the journey of fathoming the unrevealed. The beauty of age is that it cannot be held but the memories it leaves remain with us. Is an individual defined by age or age is defined by an individual is a question still under exploration and with all of its profoundness and enthralling magnificence, the authors endeavor to understand and celebrate this beautiful factor: player's age and game design.

This chapter explores the relation of age with an individual's interaction patterns with game objects. Age distinguishes behavior patterns from one age group to another. A person goes through several changes in their behavior, priorities, and emotion patterns with age. Although each individual is different from another, there are certain patterns that we see prevalent in people with similar age.

5.1   AGE AND PLAY

The role of age and experience in play is becoming an important area of study due to the changing user demographics of digital games. These demographics show a larger trend toward players becoming older. The 2016 and 2017 reports by the Entertainment Standards Association (ESA) on player demographics show an average player age of 35, up four years from 31 in the 2014 report [Entertainment Standards Association, 2014, 2016, 2017]. Not only are players aging, but new players are entering the market. As players age, their abilities and goal sets also change. The game market must cater to the needs of the players depending upon their age and experience.

Age is an enormous human factor and servicing players with concentration upon age is challenging. A player's affordance not only is impacted by age but also by other factors, such as their gameplay experience, cultural notions, gender, etc. A study on age within a population with some constant features, such as ethnicity or educational years, highlights a certain patterns among individuals, resulting from age.

Our goal is to identify player perception and gameplay commonalities among similar age groups that apply to the majority of the population. This chapter considers two age groups of players: one group comprises of teenagers and the other group has participants with an average age of 32 years. The details of the experiment and results are explained in the following sections.

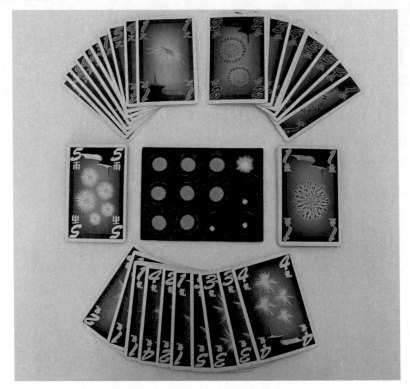

Figure 5.1: Hanabi.

5.2 PLAYTEST

This study considers the age factor and the experiment is conducted with two age groups. The tested games are Hanabi and Hive pocket, as shown in Figures 5.1 and 5.2. This experiment is inspired from a previous study upon age and game design intuitiveness [Aslam et al., 2018], where participants belonged to three age groups. Moreover, previous research work also considered the educational years along with age. During that study, it was identified that participants who were older were looking at games logically and tried to infer rules. The youngest participants mostly informed game mechanics by comparing the presented games with the games they already played.

The participants for this experiment were either students, staff, or instructors in an Information Technology University. The experiment setup is the same as mentioned in Chapter 2, presenting game without rule books. None of the participants had seen the presented games before. The participants started with filling a questionnaire as shown in Figure 5.3. The games were presented and within a specified time frame the participants had to inform the play mechanics. The responses from the participants are shown in Tables 5.1 and 5.2.

Figure 5.2: Hive pocket.

5.2.1   RATIONALE FOR SELECTING HIVE POCKET AND HANABI

This section has been adopted from Aslam et al. [2018]. Hive Pocket [Yianni, 2001] is appealing for affordance analysis since it has pieces of two colors with pictures of insects over them. The pieces do not have any text written over them but give an indication about the number of players through two colors. Other than the color, the hexagonal shape of objects gives an indication that they must be connected in a certain way. The number of pieces with a certain insect such as one Queen Bee, two Spiders, three Beetles, etc. and the name of the game, Hive Pocket, gives an indication about the importance of Queen Bee as a winning goal. Moreover, the different type of insects also give an indication about the moves or actions associated with them. Overall, Hive Pocket's objects are acting as signifiers without any text written over them. This allows us to determine to what extent participants with different ages and educational years are able to discover hints given by the objects.

Hanabi [Bauza, 2010] has been selected since it is a card game. Hanabi is also appealing for the research because it does not give a direct indication about play rules. There is no text written on the cards, just the numbers. Hanabi's cards have different colors with a number written over them. Each card has a picture of fireworks over it. During this cooperative game, the players hold

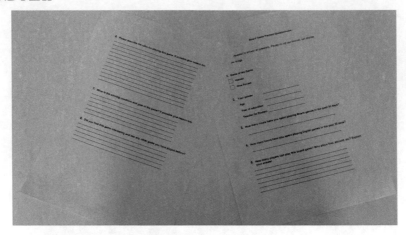

Figure 5.3: Questionnaire for playtest feedback.

cards toward other players. The different colors and the pictures of fireworks give an indication that the game is about lighting up fireworks. The tokens which are round objects with clocks and bombs printed over them give an idea of turns or time. The same question has been addressed in the study as to how much information users are able to extract from Hanabi objects.

5.3   ANALYSIS OF PARTICIPANTS' RESPONSES

5.3.1   HIVE POCKET

The Hive Pocket has been tested by two groups of participants. There were 29 participants from staff members with an average age of approximately 32 years. The students group had 58 participants and their average age was approximately 19 years.

The participants' responses from both groups demonstrate similarities of ideas, and those ideas were closer to the theme of the game. The frequently made statements from staff and student groups are presented in Table 5.1. The responses are classified together and there are 23 frequently made statements that are classified on 7 topics. The two-tailed z-score test with population proportions is used to identify if the two groups differ significantly over any category. The Null hypothesis is that no difference can be determined between the responses. Bonferroni correction to the tests determine $\alpha < 0.007301$ for each test. The statistically significant response is "*White plays first (like in chess)*," which is incorrect to the actual game mechanics. The students made this statement frequently. Those who were familiar with chess immediately related the game mechanics with chess upon seeing two sets of stones in black and white colors. The other significant response is "*Each type of insect has its own rank*," a statement the older participants (staff group) are less likely to make as compared to students. Considering the correct statements

Table 5.1: Hive Pocket: Comparison of responses from staff and student participants with z-score test for two population proportions. Statistically significant with $p < 0.007301$, as per the Bonferroni correction, highlighted.

Responses (C-Correct to Game Rules; I - Incorrect to Game Rules)	Number of Responses from Staff (29)	Number of Responses from Students (58)	p-Value
• Game is similar to Domino (I)	8	4	0.0083
• Game is similar to Majong (I)	4	3	0.16452
• Game is similar to Chess (I)	4	10	0.6818
• White plays first (like in chess) (I)	12	43	**0.00278**
• Players put stone logically, i.e., Ants cannot be placed near Beetle (I)	12	12	0.04136
• There is initial position of stones (I)	8	13	0.59612
• Stones are placed upside down (I)	2	10	0.18684
• Stone are randomly placed in hive (I)	5	4	0.13622
• The game is played by two teams as opponents (C)	23	52	0.18684
• Players cooperatively build a hive (I)	3	2	0.1936
• Bag is used in the game (I)	6	5	0.1096
• Each type of insects has its own rank (I)	5	28	**0.00496**
• Insects have different movements or abilities (C)	12	34	0.12852
• Goal is to surround opponent's stone (one or more) (I)	2	6	0.60306
• Goal is to get more stones on the table than the opponent (I)	6	10	0.69654
• Goal is to kill specific insect (Bee, Ladybug, Mosquito), one or more (C)	4	10	0.6818
• Goal is to kill all insects of enemy (I)	3	21	0.01108
• Goal is to kill enemy's Bee (C)	2	10	0.18684
• Goal is to build a hive faster than opponent (I)	2	4	1
• Goal is to collect exact type of insects (I)	4	3	0.16452
• Goal is to gain more points (I)	4	6	0.63122
• Goal is to empty hands from stones faster than opponent (I)	10	8	0.02444
• Goal is to collect exact combination (five in a row or Royal Flush) (I)	2	3	0.7414

Table 5.2: Hanabi: Comparison of responses from staff and student participants with z-score test for two population proportions. Statistically significant with $p < 0.007301$, as per the Bonferroni correction, highlighted.

Responses (C-Correct to Game Rules; I - Incorrect to Game Rules)	Number of Responses from Staff (12)	Number of Responses from Students (62)	p-Value (two-tailed)
• Maximum five players can play (C)	7	29	0.4654
• Maximum four players can play (I)	2	7	0.60306
• Clock tokens are used as timer which decide when to launch fireworks (I)	2	25	0.11876
• Black tokens end the game if all of them have been burned (C)	6	6	**0.00052**
• Cards are equally divided among players (C/I)	0	42	**0.00001**
• The game is similar to UNO (I)	1	13	0.30772
• It is similar to Durak. Each color has its value (red are stronger than greens, greens are stronger than blue) (I)	2	6	0.4777
• Each player gets a different color (I)	0	6	0.26272
• Goal is to get more cards of player's color (I)	0	13	0.08012
• Goal is to collect maximum number of different colors with high numbers (C)	3	5	0.08364
• Goal is to have highest card sum in hand (I)	1	18	0.13362
• Goal is to collect the firework of different colors from one to five (C)	4	10	0.16452
• Goal is to drop all the cards like Durak (I)	3	7	0.20408

as to game rules, the majority of staff and students considered it a competitive game. The same is true for identifying that each stone/insect has different movement patterns.

The compelling observation is the winning condition of the game, in which the Queen Bee of the opponent must be captured by surrounding it with stones. Four participants from the staff group and ten participants from the student group understood that the goal is to kill or capture a certain insect, although they mentioned not only the Queen Bee but also speculated that this insect might be Mosquito or Ladybug. The rationale appears to be that Mosquito and Ladybug are present as a single piece as Queen Bee. The accurate answer, such as capturing

Queen Bee is the goal, has been given by two participants from the staff group and ten participants from the students group. The previous study on Hive Pocket showed some participants guessing the correct winning condition but none of the participants from the youngest group (school students) could identify this [Aslam et al., 2018]. The responses from the current study are remarkably similar to the responses obtained from the previous work.

The participants related game mechanics with previously played games while guessing the rules. Those participants, who thought that game mechanics are similar to games in which the player has to get rid of tiles or cards faster than opponents to win, were excited to know the actual game rules. The participants' responses identify that Hive Pocket is intuitive to understand for broader mechanics. Suggesting exact ability or movement for each insect was difficult but the majority of the players could predict the mechanics as being insect movements or strengths. Hive has a signifier embedded in the title. Hive, as the title suggests, features the Queen Bee as an important element of the game. Furthermore, the shape of stones indicates that they can be connected or placed next to each other.

The participants' performance while guessing the rules establishes that Hive possesses a balance of signifiers such as game mechanics that become apparent as the elements in design are examined although recognizing the goal requires more attention.

Both groups of participants suggested game mechanics that are not close to the actual game rules, presenting a rather interesting setting. Game mechanics proposed by one student participant is (sic):

Units have two sides—one with insects and other is plain. Firstly all units are placed on table with insect down. Each turn, players take two units of their color and all are placed into bag. The first player takes one unit, if it is his color then it is placed on the board. Otherwise, without showing insect to the opponent, the unit is returned to initial source. The same process is for the second player. When player takes his color, he can place insect on board. The target is to surround a unit of opponent. But insects determine, which units can be placed near to each other.

Game mechanics proposed by one of the staff participant are (sic):

We can hide all insects and play like card game to match the four insects. My ordering also reveals the opponent collection due to color schema of black and vanilla. We can set two hexa-shape coins as a winner. Who ever get this one will win the game.

The participants proposed different gameplay ideas. Some of the ideas were closer to the actual rules and some of them were completely different. As illustrated above, the student is considering the bag as part of the game and also utilizing the plain side of the stone (participant referred to stones as units) as part of the play mechanics such as to hide insects from the opponent. The response from the staff participant includes the theme of hiding the stones from the opponent as they said that the game can be played like usual card games. With diverse themes, the correct rules were also suggested by some participants, as shown in Table 5.1.

Participants provided feedback about the game by informing that they find this game interesting and challenging as well. The interesting element was the possibility to connect stones in several ways. As a student expressed in feedback (sic):

This game at first seemed like Chess. Because there is a certain division into two players and units are of same rank. However the absence of board made it difficult to guess that units can move. The movement by sides is an interesting idea but still actual moves and winning condition is similar to chess. Anyway knowing this game makes me interested to play it.

Another participant from the staff group informed (sic),

Too many rules, difficult to remember.

The game design evoked interest and curiosity and participants were excited to find out the actual rules which the observer demonstrated after the playtest. However, some participants suggested the game as apparently boring and considered it a complex design especially while figuring out insect movements and abilities.

5.3.2   HANABI

Hanabi was tested by two groups of participants. One of the groups consisted of 62 students with an average age of 19 years. The other group comprised of 12 participants from the staff with an average age of approximately 34 years. The responses from the two groups of participants (staff and students) are presented in Table 5.2. The frequently made statements summed into binomial data and z-score tests for two population proportions has been conducted to identify if two populations differ significantly in some categories. There were 13 frequently made statements on 7 topics. For each test, there must be $\alpha < 0.007301$ as per the Bonferroni correction. Considering the significant results, the two groups differ in the response: *"Black token end the game if all of them have been burned."* The statement is correct according to actual rules, and it has been stated by six students and six participants from the staff group. Black tokens with a fuse pictured over them give an indication of a blast or an idea of something ending. The older group of participants observed this and the younger group (in majority) could not perceive this clue.

The other significant response is *"Cards are equally divided among players."* This statement is partially correct (as to actual rules) as cards are equally divided (either four or five cards) and the remaining deck constitutes the discard pile. Forty-two students made this statement and none of the participants from the staff group stated this. This result is critical as a number of cards dealt to each player is an important mechanic of card games. The students group majorly stated this mechanic and the staff group did not mention it.

Although not statistically significant, the winning goal of the game has been correctly stated by ten students and four staff members. The basic theme of the game, lighting up fireworks, is apparent from cards as each card has a firework picture. Furthermore, all cards are of the colors of fireworks.

However, not correct to the actual rules, younger participants (25) predicted clock tokens to be a deciding factor for fireworks display. The older group did not inform this idea in majority. The clocks on the token showing different times served as a signifier for the participants to consider them a timer. Though they do represent a notion of as to how any times a player can give the information to the other players. This proved to be difficult to predict by pictures of clocks over tokens. As the case with Hive Pocket, the current responses for Hanabi are quite similar to the responses we obtained from a previous study on Hanabi [Aslam et al., 2018].

Participants suggested game mechanics that do not differ much and contain a similar broader theme. As an example a student wrote (sic):

This game seems to be like the game "UNO." The first player places one card on the board and then one by one, players try to take place on the board by putting card with the same color or same number. The game reminds me of the Japanese event with fireworks, so may be the goal is to launch fireworks by the time. It must start at 12pm (may be). So, as the time passes, players must keep the fire on the candlewick and if one player cannot place the card required, the time increases and candlewick goes farther from firework. The goal is to launch the firework on time, no later or sooner.

A staff participant suggested gameplay mechanics as follows (sic):

A player can put cards on the basis of color or number. Players are starting with lower numbers, trying to dispose the cards from hand as fast as possible. Probably, you start with like five cards in hand, taking additional cards from the deck. Time tokens can define the rounds played, meaning that after time tokens are finished, we start using black tokens, the last one will be explosion token.

The feedback provided by students suggests Hanabi to be interesting for the majority of the participants. However, some found it not fun as it is a cooperative game. One of the feedback responses states:

Pretty unusual in terms of the fact that the game is not competitive but cooperative. Have not played anything like this before.

Another student stated (sic):

I have found the game magnificently interesting. To be honest, I did not know this kind of game before. A similar game is, when you have sticker on the head and can look at other players' stickers. To sum up, I would play this game with a great pleasure.

A participant from the staff group stated:

I find it interesting for 1–1.5 h. Have never played any game like this before.

The feedback responses were majorly similar by both groups. The majority found the game interesting because of its unusual mechanics of holding cards toward other players. Some of the participants did not consider the game much interesting as they like competitive games more or found it complex or time consuming.

5.4   DISCUSSION

The perception of participants from both groups was identical to a greater extent when they investigated game design and mechanics. We could not observe any difference based upon age of the participant. The results differ from the previous research where the participants' performance and game interaction approach was distinct among three age groups. The three groups (school students, university students, and professors) varied in age as the average ages were 14-, 22-, and 40-years-old approximately. The interaction patterns with game objects and game mechanics unfolding techniques were observed to be different. The oldest age group used a logical approach (such as defining a *conceptual model* based upon the given *signifier*, and *perceived affordance* of the game object) and tried to predict the connections among game objects. The university students were implying logic but were not as immersed in the details as professors. The youngest group of participants (school students) could only predict games while comparing them with games they have already played. The current study could not detect any difference in the participants' approach and perceptions about the design based upon age. The reason might be that two groups have an age difference of approximately ten years. Although ten years is a significant number to speculate difference in player perception and preconceived notions, the current study could not identify this.

The observations conclude that players from both groups stick to their preconceived notions of game objects. The responses suggest that participants are less likely to state any mechanics that violate their preconceived notions. This study confirms the result of the previous work that it is less likely for the participants to propose an object interaction which they have not experienced before though the object design clearly supports this interaction. Hanabi has an inverted mapping from conventional card game mechanics where players cannot see their own cards. None of the participants from either group could predict or suggest this interaction.

Identifying unconventional mechanics and object interactions is Gibson's notion of affordances where a player can experiment any action possibility supported by the design. An example of this is participants suggesting that the bag is part of the gameplay in Hive Pocket or players can cooperatively build a Hive. Such interactions can be identified in the absence of guide books as the player has the liberty to experiment. As described in Chapter 2, this playtest methodology enables game designers to identify players' perceived affordances about the object design and real affordances (as intended by the designer). Designers can embed signifiers to support and highlight real affordances and also add forcing functions to inhibit incorrect interactions.

Moreover, this does not entail, removing complexities and unpredictability that can be crucial for an adventurous play journey. This means removing confusions that are unnecessary and that design is taking into account all visceral, behavioral, and reflective levels of processing of a player's mind [Norman, 2013a]. This means considering all effects such as calmness, anxiety, hope, fear, and expectations from player's side. The readers can modify the experiment by controlling more variables such as gameplay experience and cultural background etc.

Age and play is a significant research sphere. The motivation is to have game designs that respect players of all ages and serves best to their needs and preferences. A player must be able to receive all, such as fun, education, training, or anything, the game was intended for. We believe that no single design, as a universal design, can satisfy the needs of whole game audience. The motivation for this research is to strive to achieve a design process that serves the player's needs as much as possible. The following quote from Story et al. [1998] best states our ideology of universal design:

> Universal design can be defined as the design of products and environments to be usable to the greatest extent possible by people of all ages and abilities. Universal design respects human diversity and promotes inclusion of all people in all activities of life. It is unlikely that any product or environment could ever be used by everyone under all conditions. Because of this, it may be more appropriate to consider universal design a process, rather than an achievement.

CHAPTER 6

Gender and Play

The following is the reaction of a six-year-old girl who played the board game Guess Who, in which there were 5 female characters and 19 male characters:

> "It is not only boys who are important, girls are important too. If grown ups get into thinking that girls are not important, they won't give little girls much care. Also if girls want to be a girl in *Guess Who*, they will always lose against a boy and it will be harder for them to win. I am cross about that and if you don't fix it soon, my mum could throw *Guess Who* out" [The Huffington Post, 2012].

The board game manufacturing company responded by assuring the little girl that they liked her suggestion and they will consider adding more female characters in the game. This motivates the demand for games with a *gender-neutral design*—defined as:

> A design that ensures no bias toward any gender through number, color, and appearance of objects as well as game mechanics.

Gender-neutral design is an element of the human-centered design process. The philosophy of the human-centered design focuses on making the interactions between humans and objects as desirable as possible. A desirable interaction minimizes annoyance, frustration, and confusions, and leaves a positive impression. This chapter focuses on human-centered design for games and considers gender explicitly.

Games are objects which have diversity in terms of their types, applications, and design. Entertainment Standards Association [2014] shows demographics of the computer and video gaming industry in 2014. The gaming industry has a diverse worldwide consumer base. Fifty-nine percent of Americans play video games with purchases of games divided 50% males and 50% females in the year 2014. In 2016, among the most frequent game purchasers, 60% were males, and 40% were females. In 2017, the numbers changed to 63% males and 37% females [Entertainment Standards Association, 2016, 2017]. One of the speculations for the decrease in the percentage of female purchasers from the year 2014–2017 is the failure to consider gender aspects in game designs. With the increasing popularity of the gaming industry and a number of users, significant measures are required to make games serve the purpose they are intended to for all expected consumers irrespective of gender, age, or any other human factors.

With the increased diversity of game players [Entertainment Standards Association, 2016], creating equal opportunities for all players to receive the intended benefits of the game has become an evolving issue, especially considering the gender element. This raises the question: If games are intended to cater both genders, are they being designed for both genders? Empirical testing of games is required to determine if the affordances are perceptible to both genders.

The number of female characters in a game as compared to male characters is not the only concern that highlights the issue of gender equality in games. Laydehgad [2009] refers to the morally inappropriate representation of female characters in some video games. In contrast, Sue [2016] lists board games in which female characters are not over sensualized. Along with gender portrayal, gender aspect in game design also needs to be included. Stefansdottir and Gislason [2008] define the design process as placing and patterning of any act toward a desired goal and emphasizes on the inclusion of gender aspect in design innovation processes. Furthermore, Erb [2009] and Steiner et al. [2009] highlights the significance of the inclusion of gender-sensitive approach for designing educational games to ensure equal opportunities for learning for both genders. In this context, Steiner et al. [2009] presents a model that include factors (i.e., a reason to play, competition orientation, preferences, etc.) for the consideration of gender aspects in educational video game designs.

Gender studies are crucial to detect and understand the factors that can narrow gender bias and contribute to gender equality. In this regard, Holmlid et al. [2006] discusses the complexities and challenges that might arise during gender studies. While a simple task was given to two groups, one consisting of male participants and the other consisting of female participants, the difference in actions and priorities of participants has been observed. Also, the two groups showed different attitudes toward their instructors, where male participants showed refusal in following the instructions given by female instructors this could be anticipated as a gender reaction. Jenson and De Castell [2010] in their review of 30 years of research on gender and gameplay concludes that it is time to pursue gender research without making stereotypical assumptions in the beginning.

While *gender aspect* is a broad term, our focus in this chapter which is adopted from Aslam et al. [2019b] is to identify any patterns in design (if any) that either reduce or increase the desirability of the object interaction and to further investigate if such behavior from a participant is gender-related or based upon individual preferences.

The tested game does not include objects and features in the design that have apparent inclination toward either gender. The game has been tested by two groups of participants: males and females. The intuitiveness of game design for both genders has been analyzed to:

1. study the game design from a gender equality perspective; and

2. observe how and to which extent, male and female participants perceive the intuitiveness of game design and determine which factors, if any, make them understand the game mechanics correctly or incorrectly.

6.1   METHODOLOGY

6.1.1   HUMAN-CENTERED DESIGN AND AFFORDANCES

An interesting phenomenon of human nature is perceptions, curiosity, emotions, and ways of thinking are influenced by moods and experiences [Boschi et al., 2018]. In this regard, an empirical understanding of human beings as to their potential behaviors in certain situations or interactions is indispensable for the human-centered design. Considering gender, an individual's behaviors, capabilities, needs, and other human factors are needed to understand for initiating the process of human-centered design. The goal is to identify the similarities and differences in behaviors and preferences genders. These similarities and differences must be embraced in the design process. The resulting product is a tradeoff of preferences from the recent trends of males and females' behaviors. The theory of affordances facilitates an understanding of human interactions and behaviors in terms of gender.

The research methodology adopted for gender and perceptions investigation is a playtesting process with a change in the environment of the object under testing, as detailed in Chapter 2 of this book. The change in the environment of the game is facilitated by the removal of the rule book. It is necessary for a designer to determine the people's perception of the objects and play mechanics. The goals of the testing methodology is as follows.

1. The identification of the maximum possible interpretations and perceptions about object usability. This ensures avoiding a design dimension that can lead to hurtful or unpleasant experiences on the user's side.

2. The selection of appropriate and recognizable signifiers/clues to ensure the correct usability such as letting users acquire the real affordances (usability intended by the designer) of the objects and designs.

Point (1) is referring to Gibson's notion of affordances as to all possible usability manners, regardless of if they are the object's real affordances or not. By not having guidance through rule books, players could infer any possible interaction with the game objects, which is giving a more significant set of usability and interaction patterns. This also measures the change in the game environment based upon the existence and non-existence of rules. The goal is to analyze different interpretations to filter out any undesirable aspect in the design.

Since human beings are different, and they may associate different perceptions and interpretations for the same object usability based upon their culture, gender, age, and any other human factor. Therefore, an approach toward understanding players is enabling them to play with the design with the freedom of making any move and interaction they deem possible. The demonstrated methodology is a non-competitive process. A non-competitive playtesting environment encourages players to try out various action possibilities. As there are no rules imposed, players are free to use an object in any way they think it can be used. This is a significant observation for the designers as they can see incorrect mappings of their designs and embed careful signifiers to avoid all incorrect mappings. Although knowing all sorts of perceptions toward objects and their usability is complicated and unachievable, through playtests we can approximate common beliefs, perceptions, and interpretations.

When people are interacting with the design, the interpretations of usability might differ based upon the gender of the individual. Designers can consider this for avoiding frustration and any bias toward either gender.

Point (2) highlights Norman's goal for achieving a design that naturally leads users to correct usability. The results of the playtesting process help designers to make a comparison between user's perceived affordances about the object and real affordances, such as which part of the game objects have directed players toward the correct mapping of game mechanics and which objects have led player toward the incorrect or opposite mapping of actual game mechanics.

A study of game design from Gibson's point of view of affordances is significant to understand all possible actions associated with game objects and analysis of Norman's view of affordance helps in the recognition of player's level of understanding of game mechanics as well as their interpretations of game objects. Designing games which are enjoyable for all players irrespective of gender requires the recognition of all possible interpretations and associations among game objects, for example, while playing Guess Who, a little girl perceived that female characters would mostly lose because there are more male characters in the game to choose from. Game mechanics in Guess Who does not have a gender bias. The bias exists in a set which has one trait more determinant than the other. The little girl's strategy, in this case, is probably to only select a female character, and because of this set is smaller, it makes the final determination of the character selected a more accessible pathway. The mechanic is not biased to either gender, but the theming of the game is what has introduced gender issues for the girl. Guess Who, in order to be fun must be discriminated because otherwise if it has an equal distribution of all traits, then the time to the solution will mathematically be a constant number of guesses. Therefore, in order to ensure fun, Guess Who has to be discriminatory by providing an unbalanced set of traits. At the same time, human factor's impact on the level of player's enjoyment should be investigated to take effective design decisions consequently, and game design can be modified according to the preferences and needs of the target audience.

6.2   TESTED GAME

For the investigation of gender-sensitive design[1] in board games as well as analyzing the performance of male and female participants fairly, we had listed the following criteria about the game that is to be presented to the participants. We hypothesize these criteria will provide an attempt at gender-neutral design.

1. The game objects should be gender neutral, at least in appearance.

---

[1]This study assumes a gender binary (male/female); participants gave an anonymous and self-identified response to their gender.

Figure 6.1: Goblin dice.

2. The game should not include a dominant character object that can be classified as either male (i.e., a soldier) or female (i.e., a princess).

3. The game objects should have a mix of colors—not a predominately pink pallet, which is considered to be feminine, nor a blue pallet, stereotyped as masculine.

4. The game should preferably include popular board game objects such as tiles, a board, dice, etc.

5. The basic theme of the game should be simple to understand and does not require a prior game playing experience.

Based on the criteria mentioned above, we selected Goblin Dice by Bazylevich [2015], Figure 6.1. Goblin Dice is a board game that has 22 path tiles, 1 start and 1 finish tile, 6 Goblins, 1 stone, 1 stone speed marker, and 12 dice. It is played between two to six players where Goblins are racing across a path out running the bolder trying to crush them. The winner is either the first player to reach the final tile or in the event that the stone catches up to the Goblins, sadly running them over and flattening them, the last goblin remaining.

Figure 6.2 shows the same setup before play. In the beginning, each player gets two dice. The dice are rolled into the middle of the board simultaneously. The dice can be used to perform one of two different actions. If the number on the die matches any of the numbers on the bottom of the path tile, a player can move a goblin forward. If the number on the die matches any number on the top of the path tile, a player can use special features of the tile. The game is played in rounds and continues till a goblin reaches the finish tile or all but one are flattened by the stone.

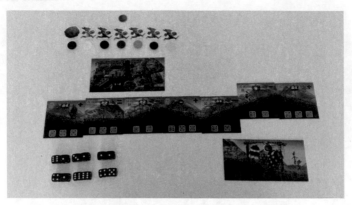

Figure 6.2: Game setup.

Figure 6.3: Game and questionnaire as presented in the test.

6.3   EXPERIMENTAL DESIGN

The study is approved by our institution's research ethics committee, and has been conducted with 80 participants, 30 females and 50 males. The participants were the student of computer science bachelors and masters program. The average age of male and female students is 20. The males have spent an average of approximately 12 hr playing board games and an average of approximately 35 hr playing digital games in the past 30 days. The females have spent an average of approximately 6 hr playing board games and the same for digital games in the past 30 days.

The game with the rule book removed and presented to each participant individually. The participants were instructed to analyze the game for 15 min and after that they had to fill out a questionnaire, (Figure 6.3). The questionnaire examines the following: (1) questions about their gameplay hours in the past 30 days and (2) game mechanics, game and winning condition, etc.

Table 6.1: Comparison of responses from male and female participants with z-score test for two population proportions. Statistically significant with $p < 0.007301$, for seven classes, as per the Bonferroni correction highlighted in bold.

Responses (C-Correct to Game Rules; I - Incorrect to Game Rules)	Number of Responses from Female Participants (30)	Number of Responses from Male Participants (50)	p-Value (two-tailed)
• The game is played by maximum six players (C)	20	36	0.6171
• The game is played by seven players (I)	0	4	0.1118
• The game is played by maximum four players (I)	7	8	0.4179
• The game is played by maximum three players (I)	3	2	0.2846
• Each player should use two dice (C)	9	14	0.8493
• Each player takes three dice (I)	5	12	0.4354
• Signs on the tiles shows how to move (C)	7	13	0.7872
• Numbers on the dice can be compared to numbers on tiles to decide action (C)	15	14	0.0477
• Goblins are running away from a stone (C)	3	10	0.2420
• Goblins are competing or playing a football game (I)	6	0	**0.0010**
• Goal is not to die from a stone (C)	1	12	0.0151
• Goal is to bring stone to the finish tile (I)	4	4	0.4413
• Player who reaches the finish tile is the winner (C)	15	26	0.8630

6.4   RESULTS AND ANALYSIS

The responses obtained from both male and female participants are summarized in Table 6.1.

6.4.1   QUANTITATIVE ANALYSIS

In order to provide a quantitative evaluation of the rules, the elements which were highlighted by multiple participants were classified together; (in all) there were 13 frequently made statements about the game on 9 topics (see Table 6.1). In order to determine the significance of the response, several common responses were summed into binomial data; the user wrote this statement or did not write the statement. Statistical tests were undertaken with a two-tailed z-score test for a population proportion. The z-score test for two population proportions is used to determine whether two populations, i.e., males and females, differ significantly on some categorical characteristic, e.g., if they said *Goblin Dice* has six players, the Null Hypothesis is that no difference can be determined between the responses. We apply a Bonferroni correction to the tests in order to remove the likelihood of the family of comparisons being in error. Given that seven topic classes are examined in order to have a $\alpha < 0.05$ for all tests, then we must have a $\alpha < 0.007301$ for each test.

One of the factors is statistically significant: females stated that *Goblins are competing or playing a football game* ($p = 0.0010$), which is not correct to the actual game mechanics. While not statistically significant, males were more likely to state that Goblins were running away from the stone ($p = 0.2420$) and state that the stone will kill the goblin ($p = 0.0151$). Taken together these findings show that females believe the stone to be in a more cooperative relationship with the goblins, whereas males believed it to be a threat or a competitive element.

In this study, the methodology removed the context of the stone which is visible via the theme, i.e., box art showing the goblins running away. Hence, males and females have reverted into their preconceptions of the mental model in terms of interactions. The default conceptual model in this case for males depicts favoring the competitiveness in-game aspects—stone as threat—and females favoring cooperation actions—stone as a football.

This leaves an open question as to how the designers of Goblin dice could have avoided this confusion. Note that the themes of the goblins are that of a sporting competition; they wear gym outfits, which fit both of the narratives of a race and a football game. The stone is the factor of confusion; the small addition of a goblin perhaps flattened by the stone behind it, or just before it looking back in horror perhaps could have clarified the purpose.

6.4.2   OBSERVATIONAL ANALYSIS

The observational analysis shows that male and female students have used the same approach while analyzing the game. The game objects were counted and similar objects were placed together; see Figure 6.4.

The maximum number of players for the tested game, Goblin Dice, have been predicted by counting the goblins. However, some of the participants also included the stone and the speed marker in the counting and assumed that there are maximum seven or eight players for Goblin Dice. Some participants also suggested three and four players because there are three dice symbols at the bottom of some path tiles, shown in Figure 6.5, and they thought that 12 dice are divided among players and each player gets 3 or 4 dice.

The correlation between the dice and the movement of goblins was understood by most of the participants as 15 out of 30 female students and 14 out of 50 male students could predict that after throwing the dice the numbers on the dice are compared to numbers on path tiles to decide further action.

The participants predicted Goblin Dice to be a competitive or a racing game in which a goblin has to reach the finish tile before others. However, a different idea suggested by six

Figure 6.4: A student analyzing the game.

Figure 6.5: Path tile with three dice at the bottom.

female participants was it to be a goblin football or a goblin soccer game in which the goblins are divided into three teams of two players and are trying to play football, rolling the stone to the finish tile. In contrast, the male participants mostly suggested that goblins are trying to run away from the stone.

Considering the stone being a football is Gibson's affordance which refers to all the action possibilities with an object. A stone can be kicked and rolled, however considering its physical properties, it is not suitable for soccer which is played using the ball. The female participants, in this case, are implying Gibson's idea of affordance as well as depicting their perceived affordance for the stone which is not an actual affordance. The goblins also look as if they are kicking a ball with one foot; this can also be the reason why females perceived it to be a football game.

There is also a difference in the way a team ballgame and a footrace interpret the game board itself. The game board suggests, somehow, what the final move would be to end the game.

Every board game includes a final state where either a player has won, or the players have accomplished a goal and play ends. Therefore, it is interesting that while male and female participants analyzed the posture of the goblins to read either running of kicking, they also found confirmation in the design of the board.

The gap between the perceived affordance and actual affordance highlights the need of a signifier to match the goal of the design. Furthermore, discussing the correct prediction and the accurate mapping of the object's intended functionality, three female and ten male participants mentioned that goblins are running away from the stone. These responses demonstrate that the stone's affordance was perceivable by these participants. The exclusion of the rule book played an important role in this study, without which it would not have been possible for someone untrained in either game design, or a form of archaeology and material culture methods, to conclusively determine possible interpretations of game objects which can significantly influence a player's experience. This can provide useful insight into the game design to add signifier and clues where needed and to remove unnecessary confusion.

The difference in male and female participants' thinking, such as stone is a destructive object, or an enemy (as anticipated by males) and stone is a fun object and an ally or a reason to win (as predicted by females), can be attributed to gender rather than merely an individual opinion. This might support the gender stereotype as females being cooperative, thinking stone as a friend and males being competitive, thinking it as a threat. However, in order to concretely prove or disprove this point, an investigation with larger sample sizes and different age groups is required.

Another insight from this experiment was that the subjects interpreted their task broadly, and had generally the same concept of what a *rule* is. All of the participants thought that they had to infer the story arc of the game, from start to finish, in order to understand the rules. There are questions they did not ask, or else they could make no conclusions worth explaining to the observers. Among those are potential rules. For example, they did not ask how many turns each player had to move. If players could only roll the dice a limited number of times, was it possible to end the game with no winners? Does the stone move as many times as the players do? Given that the playing surface is made out of multiple small boards, are there rules about constructing the board? Players also did not mention the context for playing the game. For example, none of them said the game must not be played during school, or that, for some reason, they would have to sneak it into their university dormitory. Certain conclusions can be drawn from this. The participants did not think the game was socially or politically subversive. They also considered the game to be inoffensive. Nevertheless, it is possible that male participants delighted in the idea of squishing goblins, and thought the intended audience for the game would delight in squishing goblins, too, whereas female participants might have considered injuring goblins to be offensive or sad, and decided the designers would market a nonviolent game to the intended audience.

Furthermore, the participants thought that the game had a story arc, and that arc was required to infer the rules. Rules were thought to follow logically from the story. They did not suggest arbitrary rules, like a predetermined limit to the time it takes for one complete turn.

Based upon the observation during the test and qualitative analysis, we could not see any difference in the performance of participants which can show that either group was better than the other. The difference in gender, based on the performance of participants in the test, is not apparent. The game is liked by both groups and they have expressed their desire to play the game after the test. Participants were also excited to know the actual rules and most of them requested the rule book from the observer to find out the actual rules.

A second component to the experiment would be to write two rule books—one based on the cooperative team sports suggested by some members of the female group and one based on the agonistic squishing game proposed by some of the male group. Assign participants of the experiment to play both games, and propose critical examinations of each. If they are told, the one game is designed by the male group and the other designed by the female group, does it affect their critical analysis? Under such circumstances, there would be far more detail about gender binaries and game design that the participants would reveal in the course of play and in the shared generative critical analysis. Such an experiment might actually produce a more complicated spectrum of gender-based inferences using the gaming industry's simple, binary model of male and female players.

Overall, investigating the game design results shows that the design is intuitive to lead participants in understanding that the game involves the movement of goblins and the stone. The dice and the pictures of dice on the tiles show a relation between throwing the dice and comparing the outcomes with the dice pictures on the tiles to decide further actions. There are no game objects that have misled participants toward an exact inverted mapping of the actual game mechanic. The only exceptional case was when few participants considered taking the stone to the finish line instead of goblins, as they thought it to be a goblin football game. The study with the current sample size shows that game design is intuitive for both male and female participants and does not present any bias toward either gender.

An analysis of game design based upon participants' responses suggest that there exist:

1. no bias toward any gender in-game objects to design and representation, other than the stone which was perceived as an enemy by male participants and a friendly object by female participants; and

2. the game design was intuitive for most of the participants to catch the broader theme of the game mechanics. For example, participants understood that gameplay involves goblins mobility because of the path tiles and goblins picture over the tiles (goblins appear to be running). The association between dice rolls and number on path tiles was also perceivable by most of the players.

The comments from both male and female groups who did not find the game to be fun shows it had to do with their individual preference for games and does not reflect any difference in liking or disliking of the game because of the game's aesthetic or the game's perceivable social hierarchies.

6.5   CONCLUSIONS

The motivation for this study is to make the user experience better for games for all players. The study focuses on investigating the aspect of gender in gaming such as which factors in game design introduces or removes the gender bias in the game. The game selected for testing does not include any objects that are stereotypically inclined toward either gender. The male and female groups have performed the same in guessing the broader theme of the game and game mechanics. However, a difference in performance was observed when males referred to an object in the game as a threat and females speculated it to be an ally. The difference highlights the fact that even with an attempted gender-neutral design, males and females demonstrated a different perception of the same object. This draws attention to the idea that for the inclusion of gender aspects in gaming to make them fun and create equal opportunities to acquire intended benefits of the game, it is not only object representation that is important, but how males and females perceive objects and associate meanings to them is a significant factor. The research methodology applied enables us to understand player perceptions about game objects and design. The game designers can apply this process to ensure its best suitability for the intended audience.

With the tested sample size, the study uncovers Goblin Dice to be intuitive for both male and female participants. We could see a difference in the perception of one of the objects of the game that can be speculated to be because of the gender.

Acknowledging that a single player's or a few players' interpretations of game design cannot speak for the whole game audience, still, recognition of such possible interpretations is significant for improvising game designs for clarity and exploring various play dimensions. This increases possibilities for making the accurate themes for the same game but for different audiences so that the player feels that they are a part of the system and can immerse in the play.

The playtesting method adopted for testing Goblin Dice assists game designers to test the clarity of game mechanics through game objects and to identify possible interpretations and play dimensions under the influence of human factors such as gender, age, etc. Though this does not mean removing complexities and unpredictability that are sometimes crucial for an adventurous play journey, designers must remove confusions that are unnecessary taking into account the players' needs, emotions, and expectations.

The study must be expanded with bigger sample sizes and with different age groups of participants. Moreover, other games must be tested which have objects showing the dominance of a particular gender in representation or characteristics, to compare the participant's performance in an apparent gender-neutral and biased game.

Future theoretical work is needed to engage the video game and board game industries at an international level, to examine the limited cultural assumptions about gender that informs the gaming industry's technical literature, thereby leading to constraints on the categories available for interrogation that cause gaps with social science research methodology.

CHAPTER 7

Player Perceptions of Odd-Shaped Dice for Dungeons & Dragons

While players sit around the game table, the character sheets are drawn, the pizza and pop are ready, and before them sits the adventure. However, the most important object in a tabletop game, both substantively from the playing of the game and emotionally from the storytelling experience, is the dice. No other game object demonstrates such a strong relationship, as fortune controls the fates of those players seated about the table.

Dice have historical links to both games and to religious divination. As each game was a part of ritual and its outcome was in the hands of the gods, ceremonial behavior, such as throwing of the dice, was governed by fate (known as *Cleromancy*) [Eyman, Frances, 1965]. Astragals, made from the bones of goats, have been used to cast lots since antiquity. Native American tribes used dice made from the four sided bones in their rituals, while Romans and Greeks used dice made from ivory and stone. Dice with pips and symbols made from different materials such as bone, ivory, and stone, close to the ones seen today, have been found in Egyptian burial tombs [The Metropolitan Museum of Art, y AD] remains from Native American tribes, and six-sided dice similar to our own have been in use since Greek and Roman antiquity [The Metropolitan Museum of Art, 364]. Early examples date back to 24 B.C.

This chapter has been adopted from Borodina et al. [2019]. Some of the work that has studied dice in an academic context include Isaksen et al. [2016], in which a new parameter *closeness* is defined that measures the potentiality of final score difference or closeness. De Mesentier Silva et al. [2018] present and analyze different methods to introduce stochasticity in tabletop games. The methods include dice as well as decks of cards, reshuffling the deck and conveyor belt such as putting back the drawn card with a delay. The methods were evaluated on two metrics, fairness and disparity. Furthermore, Yermolaieva and Brown [2017] examined the differences in a set of dice for the time to roll to the time to understand the roll based upon factors such as size, shape, pips used, etc. Further, they looked at the issues of readability in the dice rolls which demonstrated that a dice with markers other than pips or Arabic numerals was likely to be mistaken for a different roll based on the symbols on the die being mistaken for the placement of pips. Hence, dice which violated preconceptions were seen as more error prone to mistaken readings. Boschi et al. [2018] examine the rolls of three modified sets of dice which roll a *2D6* distribution,[1] one which was a normal pair with skewed sides and the other was a set recast to make a $2D6$ distribution. This recast set was formed into a die which summed the numbers from a $D3$ and one die with 12 sides with the faces: 1, 2, 3, 4, 4, 5, 5, 6, 6, 7, 8, and 9. It was found that players were most mistrustful of the fairness of the remolded dice, and only after playing a game of snakes and ladders with both pairs were they likely to change their minds about fairness. However, it was generally found that players enjoyed the idea of skewed/remolded dice.

We extend upon the ideas seen in these studies above by looking at a set of special purpose dice built for D&D tabletop games. The 7-die polyhedral set, used at the time as a teaching aid for mathematicians, was originally seen due to Dave Wesley's contribution to the Blackmoor campaigns [Tresca, 2011b]. Blackmoor would become an early inspiration for the first Dungeons & Dragons (D&D) [Peterson, 2012]. For this examination, we look at two 7-die polyhedral sets. Dungeons and Dragons is a fantasy role-playing game (RPG) where a dungeon master creates and moderates the adventures. Player characters have attributes that are determined by rolling the dice [Fine, 2002].

Their usually seen form is selected as Chessex®model CHX27402 Ivory w/black Marble, Polyhedral™ 7-Die Set, and via a remolded form, in the PolyHero Dice Wizard Set in Parchment & Black Ink. These sets were selected due to their similarity in the dice color and numbers in order to control for these aspects of variance and attempt to look only on the shape of the molding as the factor of interest. The motivation for this research is to investigate players' predilection for dice aesthetics as well as their opinion on die fairness when they see a usual die set and a remolded set. The remainder of the chapter details experiment design, results, and conclusions derived from the research work.

7.1 EXPERIMENTAL DESIGN

The experiment starts with playtesters filling in a questionnaire. The questionnaire consisted of questions about participant's age and average playing hours for board games over the past 30 days. Participants noted their favorite board games and if they have ever played role-playing games or not.

The observer inquired participants if they are familiar with any $D20$ system. Moreover, the participants were asked what a fair die is in their opinion. Further, the participants were shown two sets of dice; see Figures 7.2 and 7.3. Having seen the presented dice sets, the participants chose a set which they fancy aesthetically. Furthermore, the observer asked the participants if they believe either or both sets of dice are fair.

Following the questionnaire, the participants interacted with both sets of dice through playing a game based on Dungeons & Dragons [Wizards of the Coast, 2003]. A set of characters and a scenario was created (see Section 7.6), which allowed for a 10-min playtime. The game

---

[1]Common Tabletop RPG systems use a well-known shorthand for dice to be rolled of the format $xDy \pm a$ where x is the number of dice to be rolled, y is the number of sides of those dice, and a is a modifier, usually a constant or another dice roll of a different type. We will use this convention throughout the chapter.

Figure 7.1: Observer and the participant playing Dungeons & Dragons.

was played twice so that the participant gets a chance to interact with both sets of dice. The game rules were developed with consideration of participants' enjoyment. The participants are the students of a Bachelor's program in Computer Science in Russia. The game scenario was developed in a way that participants can relate with the game characters.

The Dungeons & Dragons scenario is an appeal session for student's exam. The grade has been given by the teaching assistant (TA) and students want to get a higher grade. Game characters are TAs at the university and a student. The possible Student actions include requesting the TA to reconsider the grade, showing that there is a mistake in marking, asking the TA to remark homework, or reminding the TA of their promise of bonus points for attending all lab sessions, etc. The TA actions are cancelling the exam grade appeal session, finding another mistake in the answer sheet, finding that the student cheated, etc. Rules for the play testing can be found in Section 7.6.

The play test consisted of a printed rules set, dice set, and pen for counting the health points. The observer played two games using one of the two dice sets; Figure 7.1. The participant played the student role for both games. The observer also let participants roll the die for the observer's turn to increase interaction with each die.

The observer records participants' responses and emotions such as curiosity, excitement etc. during the play session. The play session is followed by a second part of the questionnaire. The participants were asked to select the dice set, they consider fair after playing with them. Furthermore, participants informed which die from the presented sets they would prefer using for a game requiring a die roll and what is the rationale behind selecting these dice.

Figure 7.2: Dice Set 1—Chessex® model CHX27402 Ivory w/black Marble, Polyhedral™ 7-Die Set.

7.2   RESULTS AND DISCUSSION

The experiment has been conducted with 59 participants. All participants were the students of a Bachelor's program in Information Technology. The average age is 20 years. The participants have spent approximately 3 hr playing board games in the past 30 days. Some of the favorite board games mentioned by participants are Chess, Settlers of Catan, Svintus, Uno, Durak, Evolution, Alias, Dungeons and Dragons, and Arkham Horror. Thirty-seven of the participants have experience of playing role-play games. Moreover, 30 participants have informed that they are familiar with a $D20$ system. This research work has been conducted following institutional ethics guidelines. The following subsections detail results of the experimental study.

7.2.1   BEFORE THE PLAY SESSION: PLAYERS' PERCEPTION OF FAIRNESS

The participants informed their opinion of fair die. Fifty-two participants said that a die is fair if it has equal probability of all outcomes; three of the participants consider a symmetric die to be fair; and four informed that they do not think that concept of fairness holds with a die.

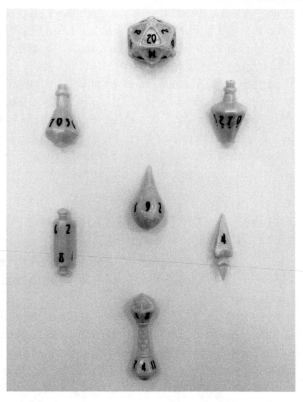

Figure 7.3: Dice Set 2—PolyHero Dice Wizard Set in Parchment & Black Ink.

While inquiring about fairness of presented sets, 32 participants considered both sets to be fair and 27 considered only Set 1 to be fair.

7.2.2 AFTER THE PLAY SESSION: PLAYERS' PERCEPTION OF FAIRNESS

The participants interacted with both sets of dice via playing Dungeons & Dragons with the observer. To ensure interaction with each die, the observer (on their turn) asked participants to roll a die on their behalf. The observer selected all dice from each set to ensure the participant interacts with all dice.

After gameplay, participants were again asked to select fair dice (in their opinion). The results indicate 19 participants considered dice Set 1 to be fair and forty considered both sets to be fair. The participants' view has been changed after interacting with the dice. Before gameplay, 32 participants considered both dice sets to be fair and after the gameplay, 40 participants considered both sets to be fair. This identify that eight participants who did not consider Set 2 to be fair changed their opinion after the gameplay. However, two participants among 40 (those who considered both sets fair) informed that they do not consider all dice in each set to be fair. One participant commented that Set 1 is completely fair, but from Set 2 only $D20$ and $D10$ are fair. The other participant said that $D20$ from both sets are fair as well as $D6$, $D10$ from Set 1 and $D12$, $D10$ from Set 2 are fair.

The observer asked participants to choose dice set which they would prefer playing with. For this question, 35 participants opted for dice Set 1. The reason for selecting Set 1 is its comfort of usage and fairness. Nineteen participants chose dice Set 2. The participants picked Set 2 because they found it more beautiful than the other set. Furthermore, 4 out of 19 participants informed that they would choose Set 2 because they are curious to use it. Five of the participants did not see and any difference and would use either of the two sets.

7.2.3   DICE DESIGN DESIRABILITY: PLAYERS' EMOTIONS AND OBJECT'S PRACTICALITY

Before gameplay, the participants were asked to select the dice set which they prefer aesthetically. In this context, 36 participants selected dice Set 1, Figure 7.2, 20 selected dice Set 2, shown in Figure 7.3, and 3 participants selected both sets of dice.

Participants who chose dice Set 1 from aesthetics point of view informed that this set appears symmetric and simple to them. Some of them called this set perfect and accurate.

After the gameplay, 35 participants selected Set 1 as dice they would prefer playing with. The 12 participants selected it because of its fairness, 16 selected because of its convenience of usage, and 7 because it looked symmetric to them.

Regarding Set 2, it was selected by 19 participants, who informed that they would play with it as it is more beautiful. Five of the participants who selected both sets informed that they do not see the difference in fairness among these dice sets and both sets appear equally attractive to them.

Examining participants' responses regarding the preference of dice for gameplay presents frequently occurring words for dice Set 2. The words include: *interesting*, *curiosity*, *cool*, and *beautiful*. The dice set which participants were not familiar with in terms of usability was preferred because of curiosity of using and uniqueness of design. The new and unusual design invoked emotions of curiosity and interest. Moreover, the design is unusual but pleasing for the participants and they considered it worth knowing more. One participant even inquired whether he can buy the Set 2 after the gameplay. This supports Norman's claim that people's preferences about things they buy is affected by emotions and sometimes emotions dominate the decision over the object's feasibility [Norman, 2004].

The responses from participants preferring Set 1 include words such as: *convenient*, *ease of rolling*, *looks familiar*, *usual*, *symmetric*, and *fair*. These responses identify that participants prioritize the practicality of the object such as a smooth roll. Furthermore, fairness of die is important for the gameplay. The participants' preference was also affected by their past experience and they felt comfortable with dice they were already familiar with.

Table 7.1: Z-score for two populations' proportions of users aesthetic perceptions of the dice. Users who had familiarity with a 7-dice set preferred the Wizard dice while new players were more likely to select the original type of dice.

	Preference for the Wizard set (Set 2)		
	Familiar (30 Participants)	Unfamiliar (29 Participants)	*p*-Value (two-tailed)
Before	15	5	**0.00782**
After	15	4	**0.00288**

Figure 7.4: *D20* from Set 1.

As we had 29 participants who had not seen both sets before (unfamiliar group), participants from unfamiliar group found normal polyhedron dice more aesthetically pleasing vs. participants who were familiar with Set 1 (30 participants) preferred the uniqueness of Set 2. The results are significant with $p < 0.05$ (Table 7.1). This supports the claim that people when familiar with something prefer unique design because of interest and curiosity. Furthermore, when people are not familiar with a particular design, they prefer a simple design when comparing simple and complex designs for the same object. The results of aesthetics preference for dice sets did not change after the gameplay.

Furthermore, the dice that caught participants' attention majorly were $D20$ from both sets and D4 from Set 2; Figures 7.4, 7.5, and 7.6. $D20$s were used more than other dice during the game, and players informed that after interacting with these dice they are certain that both $D20$s are fair.

Figure 7.5: *D20* from Set 2.

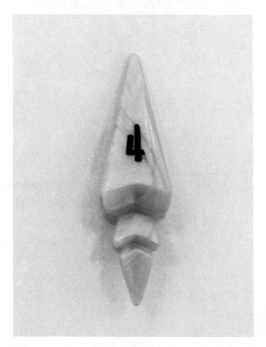

Figure 7.6: *D4* from Set 2.

Table 7.2: McNemar's test matrix and p-value for fairness of the Set 2—all participants

		After	
		Unfair	Fair
Before	Unfair	19	8
	Fair	0	32
p = 0.003906			

Table 7.3: McNemar's test matrix and p-value for fairness of the Set 2—unfamiliar with the 7-dice set

		After	
		Unfair	Fair
Before	Unfair	13	5
	Fair	0	11
p = 0.03125			

7.3 EVALUATION OF FAIRNESS

We had participants from two categories who were familiar with the usual dice from Set 1 (30 participants) and those who were not familiar with either sets (29 participants). This second set of participants are neutral while looking at both sets, and can be considered a control group for new players with those familiar representing the experienced players.

After interacting with both sets via gameplay, Table 7.2, 19 participants in total of them considered only Set 1 to be fair and 40 considered both sets to be fair. Eight out of these 40 participants changed their opinion about Set 2 after gameplay (as 32 considered both sets fair before gameplay) and also considered it fair. This shows a significant change ($p = 0.03906$) in the participants opinions over all about the fairness of dice. We notice that all participants from all groups stated that Set 1 was a set of fair dice, regardless of experience. This shows a large bias in players to see dice molded in balanced polyhedral shapes to be seen as fair and agrees with the findings in Boschi et al. [2018].

The results we see from our unfamiliar group samples, our control, as seen in Table 7.3, we find a majority of the group which is unconvinced by Set 2 being fair even after the act of gameplay. However, gameplay did have a significant effect on the belief the dice were fair after play ($p = 0.03125$).

Considering those players familiar with Set 1 but not with Set 2 as seen in Table 7.4, it is found that the number of those who are more likely to deem the dice sets fair to begin with was increased in the those who were familiar ($p = 0.01352$ on a two-tailed proportional test).

Table 7.4: McNemar's test matrix and p-value for fairness of Set 2—familiar with the 7-dice set

		After	
		Unfair	Fair
Before	Unfair	6	3
	Fair	0	21
p = 0.125			

We still find the effect of playing the game with dice increasing those who believe that Set 2 is fair, though not to a statistically significant level ($p = 0.125$). The results from familiar group of participants also indicate that Set 1 was always considered fair either alone or with Set 2.

The study gives us a useful insight into people's perception about dice design and its fairness. We identified that simplicity in the design is a factor in making people think that design is fair. An unusual design (such as wizard set, Set 2), though appreciated for its aesthetics was considered fair too but never alone. It was considered fair along with the other usual set.

The participants from familiar group considered only Set 1 to be fair in 30% of the cases (before the playtest), and the participants from unfamiliar group considered only Set 1 to be fair in 62% of cases (before the playtest). This implies that the participants who were already familiar with the *D20* system were more likely to believe that both sets are fair. We would continue the test with more samples from familiar population to test the hypothesis that a prior usability experience has an affect on perception resulting in considering the familiar sets fair.

7.4   PLAYTEST AND PLAYTESTERS' ENJOYABILITY

In context of playtesting, the common issue is gathering playtesters. Humans are an expensive resource and playtests are not compelling for the people because of time required, and the monotonous and repetitive process. This research work investigated dice design and aesthetics with consideration of playtest process's desirability for the participant.

In the process of designing playtest schema, we aimed for the process that can be interesting for the participant. In this regard, an approach is initiating a story line, a participant can relate to. The observational analysis concluded that participants were naturally immersed in the playtest. They were excited to play against observer who was playing a teaching assistant character.

The participants enjoyed exploiting their knowledge of game rules. One participant found a exploit in the game rules which allowed winning with a big score, therefore, such rules introduced excitement for the participant, and they were pleased with the feeling of discovery for the shortcut win.

The goal of the playtest is to identify participant's experience with the object in a natural play environment. The experiment detailed in this chapter successfully initiated the gameplay process for the observers to investigate research questions and for the participants to enjoy the process. This has been achieved mainly because of the story line which participants could easily connect with.

7.5   RECOMMENDATIONS FOR FUTURE WORK

The aim of this research is to identify player's perception of fairness of the dice. The research work also concentrated on the aesthetics element. The two sets chosen include usual Dungeons & Dragons dice set and a Wizard designed set.

For future work, our current experience suggests to run playtests in parallel to accommodate as many participants possible for time optimization. As a potential solution, participants can play against each other. We would expand upon the study by increasing sample size from both group of participants: familiar and unfamiliar with the dice sets to investigate trends toward fairness perception in case of prior familiarity and unfamiliarity.

There is also a large missing area of research work from an cultural anthropology perspective to examine the roll/role of dice.

7.6   GAME RULES

There are two players in the game: one player is the participant of our research and the other player is the observer. The participant plays the "Student" character and the observer plays the "Teaching Assistant (TA)" character. The rules were explained to the participant in the following manner:

You are about to play a game and the rules are the following. You are an unlucky Student who has always been on the lowest scholarship every single semester. Presently, there is one particular subject that you have a lot of trouble with and you hope that if you show up on the appeal session, you will get a raise in your grade. You come to the appeal session and face the TA, who is not willing to give you grades that easily.

In the list shown below, you can see your HEALTH characteristic and ACTIONS you can perform. You will play two times and each time there will be several rounds. Your goal is to make the TA lose all their HEALTH points before you do.

Each round is as follows.
Your move:
State the ACTION you want to perform.
Roll $D20$.

1. If the result is greater or equal to the DIFFICULTY of the action then roll the DAMAGE die. The result will lower the HEALTH of your opponent according to the number rolled (and possibly the additions).

2. If not, skip your turn.

TA's move:

The TA (me in this case) will state their ACTION.
You will roll $D20$ for them.

1. If the result is greater or equal to the cost of the action, you will roll the DAMAGE die, lower your HEALTH according to the number rolled. Moreover, write your new health near HEALTH characteristic.

2. If not, the TA skips the turn.

This will continue until one of us loses all our HEALTH.

7.7   HEALTH CHARACTERISTICS AND ACTIONS

7.7.1   TEACHING ASSISTANT (TA)

HEALTH: 20
ACTIONS
Magic Disappearance: The appeal session is suddenly canceled. DIFFICULTY: 14+. DAMAGE: $D10$. Can only be played once.
Reverse Appeal: The TA found another mistake in the Student's work. DIFFICULTY: 9+. DAMAGE: $D6$
Cheating Suspected: The TA thinks that the answers of the Student look quite similar to some other student's answers and they already saw it somewhere else. DIFFICULTY: 12+. DAMAGE: $D6 + 4$.
Unfamiliar Face: The TA does not remember seeing the Student at all. DIFFICULTY: 10+. DAMAGE: $D4 + D8$.
MOODLE: MOODLE just does not work and there is nothing the TA could do. DIFFICULTY: 7+. DAMAGE: $D6 + 2$.

7.7.2   STUDENT

HEALTH: 25
ACTIONS
The Begging: Tell that, this is the only subject that you failed, so you really need these couple of points. DIFFICULTY: 16+. DAMAGE: No damage next move. Can only be played once.
Mistake Correction: The student thinks their answer is right and there is a mistake in marking. DIFFICULTY: 11+. DAMAGE: $D6$
"Can you check my homework again?": The Student tries to change the subject. DIFFICULTY: 12+. DAMAGE: $D10$

Really Familiar Face: The TA confuses the Student with someone and suddenly remembers that they promised this Student bonus points for attending all the labs. DIFFICULTY: 14+. DAMAGE: $D12$.

The Copy: The Student shows another Student's exam sheet with the same answers but with a different grade. DIFFICULTY: 7+. DAMAGE: $D8$

CHAPTER 8

Dice Design and Player Preferences for Colors and Contrast

Colors are associated with perceptions, emotions, aesthetics, and usability. Colors have the power to introduce diversity in many shapes and forms. Colors invoke experience and emotion. A glorious day is a blue sky and an unfulfilled wish is a black hole. It is the millions of shades spread all over the sky that make a sunset beautiful. Colors depict all moods and embrace emotion. Colors are joy and sadness, fantasy and reality, longings and achievements. Everything has colors: adventures, perceptions, hopes, failures, impressions, and experiences. Humans, by nature, associate colors with life. Life, in brief, is a transition of color.

> Colors speak all languages.
> *Joseph Addison, Spectator. No 416. Friday, June 27, 1712*

Recognizing the enormous predomination of colors and their impact on our well being, this chapter, which is adopted from Aslam et al. [2019a], is an effort toward understanding an individual's preference for game object colors. The question under investigation is how much a player's preference of colors in general is correlated to their preference for game object colors aesthetically. Though colors are also introduced to increase design clarity for users, our major focus is on identifying preference correlation between general preferences and game object color predilections.

Colors can speak for the design usability as much as the designer wants them to illustrate. Cultural aspects must be taken into consideration while determining colors to represent a certain dimension of design usability. The focus of this chapter is on players' color preferences from an aesthetic point of view. The question under investigation is how much a player's preference of colors in general (such as in everyday life) is related to their preferences of colors with game objects. Games are artistic objects with aesthetic concerns which are best evaluated by human subjects testing. Anything can be a subject of aesthetics and color is undoubtedly a significant component of art. Though color preference is a subjective matter, research on colors and emotions still have proved that common patterns exist between color preferences and appreciation in particular contexts.

There has been enormous research on color psychology and readers can find plenty of compelling investigations on this topic. A few examples are: Lee et al. [2009] studied color preferences among different age groups; Mohebbi [2014] investigated color preference in children based upon their gender; and Hanafy and Sanad [2015] investigated color preferences between groups of different educational backgrounds. Color psychology is significant to dive into in order to find correlations among color choices and human factors, affecting an individual's preference among colors. Colors have the capacity to be a visual language [Muljosumarto, 2018]. Though in-depth color psychology is beyond the scope of this research, we present a rather simple experiment that investigates if the individual's preference for certain colors also applies to a game object. In this chapter, dice have been concentrated upon for analyzing color preference of game objects.

Dice has been a significant element of games. The six-sided die has existed from antiquity and bone-based pipped examples dating from the Roman and Byzantine 1st–3rd century (see Figure 8.1), are little changed even in the 9th–10th century (see Figure 8.2). These dice could easily be used in the games of today, though not to the same levels of manufacturing quality to ensure the distribution.

Dice manufactured in a variety of shapes, colors, and sizes are fascinating for the game audience. Research on current trends of players' perception about the fairness of dice as they see dice of unusual shapes has been presented by Boschi et al. [2018]. The results demonstrate that participants' preferences about dice usage are influenced by their past usability experience. The research participants considered those dice fair which they have seen or have interacted with previously. However, they also showed curiosity and interest in interacting with the dice, which appeared unusual in design and aesthetics. Dice have also been examined in an academic context in regards to their usability for contrast and errors [Yermolaieva and Brown, 2017].

This user study evaluates the impact of colors on usability mechanics and aims to understand players' preferences of colors for game objects, aesthetically. The chosen object for this study is dice as they are familiar objects with clear usability and mechanics. We aim to identify how much an individual's choice for colors in daily life is inflicted upon their choice for colors in game objects. Furthermore, we also investigate color contrast and die readability experience. In case a player's preference for game object colors is synchronized with their color preferences in general, game designers can utilize significant findings from color psychology research to deduce their audiences' recent trends for color predilections.

The experimental study elaborated in the following section examines the correlation of an individual's likeness pattern of colors in general with their selection of colored dice to interact with.

8.1 EXPERIMENTS

The following sections describe the research methodology. All participants began by answering a short questionnaire about their age and gender.

Figure 8.1: Bone dice and knuckle bones from Byzantine Rome c. 1st–3rd century—Istanbul Archaeology Museums, Turkey—photo by authors.

Figure 8.2: Dice from the Sarkel Fortress in Rostov Oblast c. 9th–10th century—State Hermitage Museum, St. Petersburg, Russia—photo by authors.

8.1.1 COLOR BLINDNESS TEST

The Ishihara color blindness test was chosen to identify visual impairment [Ishihara, 1987]. This test was chosen as it is known to produce accurate results as well as does not require any particular equipment and is easy to conduct online. The participants were subjected to online "Ishihara color blindness test" that contains 38 pseudo-isochromatic plates, each of them showing either a number or some lines [Colblindor, 2018]. According to what the individual can and cannot

Figure 8.3: First phase—papers for sorting in nine colors.

see, the test gives feedback on the degree of their red-green color vision deficiency. The test gives result as *none*, *weak*, *moderate*, and *strong*. The analysis considered samples which indicated the *none* result, meaning, no red-green color blindness. The color-blind samples in the data are treated as outliers for this research.

8.1.2   COLOR PREFERENCE

The 68 participants (average age is 22 years approximately) were the students and staff in an Information Technology university. The participants were shown nine different colors on paper; Figure 8.3. The colors include red, pink, yellow, white, green, light blue, dark blue, violet, and black. The participants had to sort the papers from the most preferable color to the least preferable in this set. Once the paper colors were sorted according to the participant's preference, transparent dice of nine colors (same colors as shown with papers) were presented to the participant who was asked to sort them from the most preferable color to the least preferable (see Figure 8.4).

8.1.3   COLOR CONTRAST AND DICE SORTING

After identifying participants' preference of colors in general and to dice, the next step involves identifying if different background colors of the dice with an apparently readable contrast affect the dice readability or not.

For this, the participants were asked to sort four transparent dice of five colors including red, blue, purple, yellow, and green in ascending order; see Figure 8.5. The sorting time was

Figure 8.4: First phase—the transparent dice in nine colors (same as paper colors).

Figure 8.5: Second phase—sorting transparent dice in ascending/descending order.

recorded by the timer and is done two times for each color for more precision. Further, the participants were given opaque dice of two colors—red and white—for sorting in ascending order; see Figure 8.6.

Figure 8.6: Second phase—sorting opaque dice in ascending/descending order.

Figure 8.7: Unusual color contrast dice for playing knockout.

Figure 8.8: Control die.

8.1.4   KNOCKOUT GAME AND DIE READABILITY

The third part of the study includes seven dice with unusual color contrasts and a usual white six-sided die, as shown in Figures 8.7 and 8.8. Thirty participants with an average age of approximately 21 years were selected for the readability experiment. The participants were students and have spent approximately 7 hr playing video games and 3 hr playing board games on average in the past 30 days. Furthermore, 13 participants informed that they prefer video games over board games and 7 participants prefer board games. Ten participants showed neutral response and like playing both digital and non-digital games. The experiment includes playing a knockout game with the observer with each die from the set in Figure 8.7.

The observer selected a die, one by one, from eight dice and played the knockout game. The hypothesis under test is that contrast is the main reason that affects the readability of the die. The die in Figure 8.8 is the control die as it is commonly used and the most familiar one among all dice presented. The game starts with players choosing a number from one to six. The player rolls the die and *reads the outcome* once the die stops rolling. The one who rolls the opponents number first is the winner. During the game, the observer is counting as to how many times the participant got confused (such as, had to concentrate more while reading) and how many times they read the outcome incorrect.

8.2   EXPERIMENT RESULTS

We have the hypothesis that selections of color preference of an individual should be linked to their color preferences for dice. In order to examine this hypothesis in the study, we examine the color arrangements for the number of differences. We will assume that given the extremely short period of time between the presentation of the slips of paper and of the dice, the emotional stability was not a factor which perturbed the ordering. In psychological tests such as Lüscher color test, the participants, when presented with colors, are asked to concentrate on colors in front of them and not think of colors that would suit them in dresses or they would like to see in other things. In our study, we asked participants to choose their preferred colors. The preference for a particular color might result from their likeness of this color because of other factors such as this is their favorite color for clothes etc. The hypothesis under test is how much an individual's choice of colors, in general, is reflected upon their choice for selecting dice from most favorite to least favorite colors.

For the difference calculation, we need to examine what is a difference in the arrangement so we can place a measure. We assume that people may forget a color, add a new color, substitute one color for another. This is modeled well by the edit distance (also known as the Levenshtein distance) which looks at the minimal number of insertions, deletions, or substitutions to turn one string into another. We can also examine an extended version which allows for elements to be transposed or swapped. Now we have a measure on the difference, for each of the players, we find the difference between the string and produce a histogram. What we would expect to see if people's selection of colors, in general, is close to their selections on the dice is a histogram heavily weighted to low numbers of edits being required to turn one string into another, that disruption is very unlikely.

Examining the histograms in Figures 8.9 and 8.10, we come to a result which violates our hypothesis; rather than seeing a high peak at the low end of perturbations, we see a peak at about the middle. Therefore, color preference alone does not account for dice color preference. But we should look at where the perturbation is happening in the string. Perhaps it is that people care

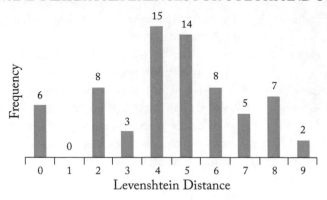

Figure 8.9: Levenshtein distance calculated without swapping.

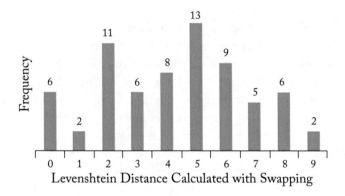

Figure 8.10: Levenshtein distance calculated with swapping.

more about their first choices of die color being their favorite color, or some early subset has importance.

However, examining the first choice of color, only 31 out of 68 participants maintained their first color choice between both paper and dice. The first three color preferences both for paper and dice colors were examined and Figure 8.11 shows which colors were replaced from the first three while participants informed their preferences for dice colors. The new colors that were introduced in the first three color sets are shown in Figure 8.12. With the introduction of new colors, we could not see one dominant trend such as either moving to a darker color from lighter ones or vice versa was significantly prominent than the other.

Based on these results, the color preference seems to have a poor relation to dice color preference. Given that these dice are of the same size, shape, weight, transparency, manufacturing company, etc. There is another factor in the minds of the participants rather than just enjoyment of colors. We believe it is usability.

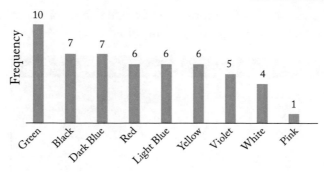

Figure 8.11: Replaced colors from first three preferences.

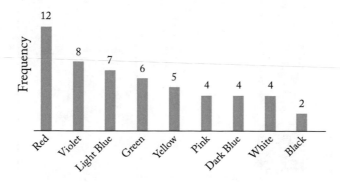

Figure 8.12: New colors that are introduced while selecting favorite dice.

8.2.1   RESULTS OF THE SORTING

We took the average time of the sorting for all 68 participants for each of the 5 types of dice and looked at a comparison via a one way repeated measures ANOVA, which found no statistically significant difference between the dice sorting times for the 5 transparent sets of dice or the opaque dice sets.

8.2.2   KNOCKOUT GAME AND READABILITY ERRORS

The results of the knockout game and dice readability experiment are demonstrated in Table 8.1. The maximum number of error, i.e., incorrect reading is two. The transparent die with white dots (3rd row in Table 8.1) caused the most errors (two) and 26 times led participants to focus more on reading the outcome. The orange die with black dots caused one reading error. It is speculated to be because of the brightness of the orange shade.

The participants provided feedback on the ease of readability and informed that the control die, the usual white die with black dots, Figure 8.8, is the most comfortable to read in comparison to other dice presented. Participants considered transparent die with white dots (as shown

Table 8.1: Dice with error and confusion rate—error refers to incorrect reading and confusion refers to correct reading but required more concentration

Die	# of Errors	# of Confusions
	0	0
	1	0
	2	26
	0	0
	0	0
	0	2
	0	0
	0	2

in Table 8.1, 3rd row), the most inconvenient to read. In context of color desirability, transparent grey dice (both with white and red dots, 4th and 8th row in Table 8.1) were most fancied. The contrast of grey and white, as well as grey and red, including the transparency of the dice, appeared fashionable and neat to participants and therefore caught attention. Though the transparent white die with white dots was the most difficult to read, participants were enthusiastic about it and called it exotic and interesting.

The analysis needs extended testing to discern a particular color contrast as providing a hindrance to smooth reading of die roll. The results obtained so far clarify the significance of color contrast selection for game designs. Dice being a prominent article in games requires consideration upon design dimensions to avoid unnecessary complications with usability during gameplay.

8.3   SOME COMMENTS FOR THE GAME DESIGNERS

This chapter focuses on colors for game articles. The die has been selected for the investigation as it comes in variety of colors and is easy to interact with. As mentioned earlier, *colors* is a huge domain to understand as they exist in our inner and outer world. The study presented here uses particular steps to analyze participant's interaction with the colored die based upon our hypothesis. Designers while investigating color impacts can apply any methodology for the test. A simple affair such as a color contrast has the capacity to induce frustration into player's interaction with the game design.

Colors are significant from both dimensions of *possible error introduction* and *making an interaction pleasant*. It is a trade-off as most aesthetically appealing design and colors might not give a smooth usability interaction. The decision lies with a designer's priority and goals for the game and its players.

CHAPTER 9

Conclusions

Players—these are users, consumers, customers, passionate fans—in game design will occupy your mind from sunrise to sunset. The act of designing the game can certainly be a passion project but the player is ignored. User-centered design has become, in some domains, a buzz word—yet the intent remain valuable, a focus on those who use the product. This book primarily examines analog games as its examples. We have done this for three reasons: (1) lack of studies in analog games; (2) the diversity in interfaces in analog games; and (3) the design process of analog games is transferable skill set to digital games.

The first reason is of primary interest to the academically inclined. The study of games is well recognized by its practitioners as being a significant field and while the domain is expanding from an academic perspective, much of the focus of inquiry has been made on digital games. Primarily, most of the practitioners are coming from the fields of Computer Science (Brown), Computer Engineering (Aslam) with a focus on Artificial Intelligence and Mathematics. This leaves analog games as a fertile ground for new research and we hope this book serves as an invitation for future colleagues. The second reason is the diversity of interfaces. Digital games are bound to the hardware, and there is incentive for designers of games to standardize and develop for the PC and console-markets which means acceptance of the input methods. This reduces many issues of design from "How does C do X in my game?" to "How does C do X in my game using the console's standardized controller." While still an interesting question, the design of mappings on these controller is defined. In order to demonstrate the techniques in the book, we take the focus off the standardized hardware and look at the design of an interface. The other cases of focus are the detailed issues without the extra layer of the controller.

Finally, the early design process of games is a story of digitizing analog examples. Magic the Gathering is not only the basis for its digital implementation, but is the foundational basis for Hearthstone. Collectable Card Games (CCG) are also a demonstrable precursor of lootbox systems in games monetization. Focusing on the design of the analog example allows us to focus on the salient features of design, quickly make changes, and playtest with minimal setup costs.

A design process with the intent of less dependency on rule books revolves around player perceptions and affordances. This book is a guide to understanding design dimensions according to player preferences and affordances. Each chapter demonstrates a playtest process to investigate different design aspects as well as player interests. The reader will learn various techniques of playtest and understand how the playtest methodology must be modified according to research questions. Each playtest process outlines a different story to inspect various facets of game design.

The joy of working on such methods comes from the people you meet and the understanding you obtain of players. Games are a social process more than just a purely mathematical one. The role of the academic in this field is to act as a keen observer of humans engaged in play. Playtesting is a challenging process starting from the moment game objects are exposed to players till analyzing their very responses. The challenge is to understand the rational behind player's story. The feedback from the player starts as soon as they are exposed to game objects. Players express feelings and show appreciation of design aspects because it reminds them of a childhood fable, gives them a sense of power, allows them to personify a character they truly adore, and much more.

Our most astonishing experience of this playtest journey was understanding how simple things are under the cover of complex patterns. Reading through the player stories and feedback, we got acquainted with fascinating human beings. The playtest methodology uncovered, not just what players like or dislike but why they have such preferences. Small details in design influence play experience. With a lot more remaining to explore, we realize it is usually small things that bring joy to players.

We handover this book to the readers with utmost bliss and we wish to hear many new stories of players and gameplay developed on this work.

APPENDIX A

Statistics

We include this short appendix in order to examine some of the statistical tests used in this work. This is for informational purposes and is not to stand in for any statistical training for a proper design of experiments. The tests used met with our experimental designs and before using a statistical test it is best to ensure that it matches with the data, i.e., talk to a statistician during your Design of Experiments.

A.1   COMMON ERRORS

A.1.1   LACK OF VARIANCE

As the field of player testing grows, these tests become more relevant for publications. Often, we see a number of claims rejected or which should have been rejected due to a lack of tests. Perhaps the most seen error is the presentation of a mean value without any display of variance or test.

Assume, for example, you have a test with ten participants for each criteria playing some game with both positive and negative scores, and all these populations having a mean of 0. Can we say that the testing cases are equivalent? No.

This could mean almost anything in terms of the difference in variance. It could be that everyone of the ten scored 0; no variation. It could be that everyone has a bell-shaped distribution about 0 with a standard deviation of 1. It could be that half the population person scored 1 million points and half lost 1 million.

So, not only is the mean value of a criteria important, but so is the variation of the scores. This also calls into question the repeatably of the study, low variance with a large population in a study is preferred as it shows the results are stable to other unknown factors in the testing. Simply putting just a mean for a small population is a good way to have your research claims rejected.

A.1.2   LACK OF HYPOTHESIS TESTING

Perhaps most pervasive is the lack of testing a hypothesis and just a declaration that since the means are different, then a difference has occurred. But, note that as we have seen above, just because a mean is different does not easily imply that the distributions are not similar. So, without looking at the variance we have no idea if this difference in means is due to random chance (the variation in the outcomes) or is because there is an effect from the change. Hypothesis testing is what gives us the confidence in such claims, usually at a rate of risk of 1 in 20 or a p-value < 0.5.

By not having the hypothesis test, we cannot make claims like A is better for our players than B, with any degree of certainty, and is likely to see a research claim rejected.

A.2   TESTS USED IN THIS BOOK

We briefly give some details on the statistical tests used in this book.

A.3   T/Z-TESTS

A.3.1   T/Z-TEST FOR TWO POPULATION PROPORTIONS

Perhaps before now you have seen the t/z-score test against a normal variable. However, the population proportion test examines the same factor based on a proportion of a population instead.

Given the number of positive tests, p_1 and p_2, and the respective number of samples of each group n_1 and n_2. We first construct the proportions $\hat{p}_1 = p_1/n_1$ and $\hat{p}_2 = p_2/n_2$.

Then we need the overall sample proportion, that is how many occurrences of a positive result happened over all tests, $\hat{p} = \frac{p_1+p_2}{n_1+n_2}$.

This allows us to construct a final testing score of:

$$Z = \frac{\hat{p}_1 - \hat{p}_2}{\sqrt{\hat{p}\left(1 - \hat{p}\right)\left(\frac{1}{n_1} + \frac{1}{n_2}\right)}}.$$

Which is no doubt a familiar construction to the normal t/z-score tests.

A.3.2   BONFERRONI CORRECTION

If we want to have a number of tests on the same set of data, then each test that we make has more opportunity to introduce errors in our confidence that the test works. The *Bonferroni Correction* distributes the risk of error across a number of tests in order to ensure that the total error is less than the expectation. For example, if we are willing to accept 1 in 20 risk over N tests on the same data, then we would need to distribute to each individual test an acceptable error of $1/20N$. Then the sum of these errors could never be more than 1 in 20. Note that this is a gross correction, there are more precise methods of this calculation when more is known about the distributions, but in most cases it works well.

A.3.3   ANOVA

Analysis of variance (ANOVA) allows for the evaluation of the variance on a group of settings $(N > 2)$ and examines if there are differences which we should explore further. It will give an

Table A.1: Contingency table

	Test 2 Pass	Test 2 Fail	Sum
Test 1 Pass	a	b	$a+b$
Test 1 Fail	c	d	$c+d$
Sum	$a+c$	$b+d$	$n=a+b+c+d$

alert if it believes there is a significant difference in the means between some of the settings. This can then be followed up with further tests, such as the t/z-test to find the differing means.

A.3.4   MCNEMAR'S TEST

This is a test for numerical data which is paired that does not conform to a bell distribution but to a binary. For example, we have a trait which is present or not present. A contingency table (Table A.1) is first created which shows the before and after of the trait.

Given such a table we are most interested in the discordant—that is, those who passed the test before and failed after, b, or those who failed before and now have passed c.

We can then look at the χ^2 table, where $\chi^2 = \frac{(b-c)^2}{b+c}$. When these values are small, $b + c < 25$, there is an exact test using the binomial distribution, which we will not explain here.

Bibliography

Aslam, H., Brown, J. A., and Baba, E. (2019a). Dice design respecting player preference for colours and contrast. In *Proc. of the 14th International Conference on the Foundations of Digital Games*. ACM. DOI: 10.1145/3337722.3342237 73

Aslam, H., Brown, J. A., Nikolaev, E., and Reading, E. (2019b). Gender and play in goblin dice. In *20th International Conference on Intelligent Games and Simulation*, pp. 19–25. 46

Aslam, H., Brown, J. A., and Reading, E. (2018). Player age and affordance theory in game design. In *19th International Conference on Intelligent Games and Simulation, GAME-ON*, pp. 27–34, Eurosis. 12, 34, 35, 39, 41

Bauza, A. (2010). *Hanabi*. R&R Games Inc. 35

Bazylevich, M. (2015). *Goblin Dice*. Game [Table Top], Foxgames and GaGa Games. Poland. 49

Borodina, K., Aslam, H., and Brown, J. A. (2019). You have my sword; and my bow; and my axe: Player perceptions of odd shaped dice for dungeons and dragons. In *Proc. of the 14th International Conference on the Foundations of Digital Games*. ACM. DOI: 10.1145/3337722.3342236 59

Boschi, F., Aslam, H., and Brown, J. A. (2018). Player perceptions of fairness in oddly shaped dice. In *Proc. of the 13th International Conference on the Foundations of Digital Games, FDG'18*, pp. 58:1–58:5, ACM, New York. DOI: 10.1145/3235765.3236389 47, 59, 67, 74

Brown, J. A., Aslam, H., Makhmutov, M., and Succi, G. (2019). Intuitive rules design evaluation methods and case study. In Guzdial, M., Osborn, J. C., and Snodgrass, S., Eds., *Proc. of the 2nd Workshop on Knowledge Extraction from Games Co-Located with 33rd AAAI Conference on Artificial Intelligence, KEG@AAAI*, Honolulu, Hawaii, January 27, volume 2313 of *CEUR Workshop Proceedings*, pp. 35–42, CEUR-WS.org. 12

Browne, C. (2015). Embed the rules. *Game and Puzzle Design*, 1(1):60–70. 12

Castle, K. (1998). Children's rule knowledge in invented games. *Journal of Researches into Childhood Education*, 12(2):197–209. DOI: 10.1080/02568549809594884 15

Colblindor (2006–2018). Ishihara 38 plates CVD test. Online accessed, April 2, 2019. 75

Daviau, R. (2011). Design intuitively. In *The Kobold Guide to Board game design*, pp. 42–49, Open Design, Kirkland, WA. 10

Dawson, D., Borin, P., Meadows, K., Britnell, J., Olsen, K., and McIntryre, G. (2014). The impact of the instructional skills workshop on faculty approaches to teaching. *Technical Report*, Higher Education Quality Council of Ontario. 16

De Mesentier Silva, F., Salge, C., Isaksen, A., Togelius, J., and Nealen, A. (2018). Drawing without replacement as a game mechanic. In *Proc. of the 13th International Conference on the Foundations of Digital Games*, pp. 57, ACM. DOI: 10.1145/3235765.3235822 59

Entertainment Standards Association (2014). 2014 essential facts about the computer and video games industry. *Technical Report*. 33, 45

Entertainment Standards Association (2016). 2016 essential facts about the computer and video games industry. *Technical Report*. 33, 45

Entertainment Standards Association (2017). 2017 essential facts about the computer and video games industry. *Technical Report*. 33, 45

Erb, U. (2009). Reflections on gender aspects of designing an educational PC-game exemplified by a project in cooperation with the German maritime museum bremerhaven. In *5th European Symposium on Gender and ICT*, University of Bremen, German. 46

Eyman, Frances (1965). American Indian gaming arrows and stick-dice. *Expedition Magazine 7.4*, n. pag. Expedition Magazine. Penn Museum, March 18, 2020. http://www.penn.mus eum/sites/expedition/?p=1038 59

Fine, G. A. (2002). *Shared fantasy: Role Playing Games as Social Worlds*. University of Chicago Press. DOI: 10.2307/539941 60

Gibson, J. J. (2014). *The Ecological Approach to Visual Perception: Classic Edition*. Psychology Press. DOI: 10.4324/9781315740218 5, 14

Gibson, J. J. (1986). The theory of affordances [Chapter 8 extract: The ecological approach to visual perception]. In *The Ecological Approach to Visual Perception: Classic Edition*. L. Erlbaum, London. DOI: 10.4324/9781315740218 5

Golomb, C. and Kuersten, R. (1996). On the transition from pretence play to reality: What are the rules of the game? *British Journal of Developmental Psychology*, 14:203–217. DOI: 10.1111/j.2044-835x.1996.tb00702.x 15

Hanafy, I. M. and Sanad, R. (2015). Colour preferences according to educational background. *Procedia-Social and Behavioral Sciences*, 205:437–444. DOI: 10.1016/j.sbspro.2015.09.034 74

Holmlid, S., Montaño, F., Johansson, K., Datavetenskap, N. D.-O. I., and Datavetenskap, N. D.-O. I. (2006). Gender and design: Issues in design processes. *Proc. from Women in Information Technology*, University of Salford. 46

Isaksen, A., Holmgård, C., Togelius, J., and Nealen, A. (2016). Characterising score distributions in dice games. *Game and Puzzle Design*, 2(1). 59

Ishihara, S. (1987). *Test for Colour-Blindness*. Kanehara Tokyo, Japan. 75

Jenson, J. and De Castell, S. (2010). Gender, simulation, and gaming: Research review and redirections. *Simulation and Gaming*, 41(1):51–71. DOI: 10.1177/1046878109353473 46

Laydehgad (2009). Gender in game design. Online accessed, April 2, 2017. https://laydeh gad.wordpress.com/ 46

Lee, W.-Y., Gong, S.-M., and Leung, C.-Y. (2009). Is color preference affected by age difference. *International Association of Societies of Design Research*, pp. 1837–1846. 74

Linderoth, J. (2013). Beyond the digital divide: An ecological approach to gameplay. *Transactions of the Digital Games Research Association*, 1(1). DOI: 10.26503/todigra.v1i1.9 12

Macpherson, A. (2012). The instructional skills workshop as a transformative learning process. Ph.D. Thesis, Simon Fraser University. DOI: 10.13140/2.1.4296.7685 16

Mateas, M. (2001). A preliminary poetics for interactive drama and games. *Digital Creativity*, 12(3):140–152. DOI: 10.1076/digc.12.3.140.3224 12

Mohebbi, M. (2014). Investigating the gender-based colour preference in children. *Procedia-Social and Behavioral Sciences*, 112:827–831. DOI: 10.1016/j.sbspro.2014.01.1238 74

Muljosumarto, C. (2018). A case study color as a visual language: Focused on TV commercial. *Nirmana*, 17(1):1–9. DOI: 10.9744/nirmana.17.1.1-9 74

Norman, D. (2004). *Emotional Design: Why we love (or Hate) Everyday Things*. Basic Books, New York. 6, 32, 64

Norman, D. (2013a). Chapter 2: The psychology of everyday actions. In *The Design of Everyday Things*, pp. 37–73, Basic Books, expanded edition. 6, 7, 42

Norman, D. (2013b). *The Design of Everyday Things*. Basic Books, expanded edition. DOI: 10.15358/9783800648108 6, 8, 9, 13

Peterson, J. (2012). *Playing at the World*. Unreason Press. 15, 16, 60

Stanfield, R. B. (2000). *The Art of Focused Conversation: 100 Ways to Access Group Wisdom in the Workplace*. New Society Publishers. 15

Stefansdottir, S. and Gislason, H. (2008). Design innovation for gender equality. http://ww
w.soleystefans.com/wp-content/uploads/2011/02/DIG-Equality_web.pdf 46

Steiner, C. M., Kickmeier-Rust, M. D., and Albert, D. (2009). Little big difference: Gender aspects and gender-based adaptation in educational games. In *International Conference on Technologies for E-Learning and Digital Entertainment*, pp. 150–161, Springer. DOI: 10.1007/978-3-642-03364-3_20 46

Story, M. F., Mueller, J. L., and Mace, R. L. (1998). *The Universal Design File: Designing for People of All Ages and Abilities*. 43

Suchman, L. (1993). *Response to Vera and Simon's Situated Action: A Symbolic Interpretation*. DOI: 10.1207/s15516709cog1701_5 12

Sue, T. M. (2016). Climbing chutes and ladders, the positive representation of women in board games. Online accessed, April 2, 2017. https://www.themarysue.com/women-in-board-games/ 46

Suits, B. (1978). *The Grasshopper: Games, Life and Utopia*. University of Toronto Press. DOI: 10.5840/teachphil19814114 15

The Huffington Post (2012). Jennifer O'Connell, mom, and 6-year-old daughter ask Hasbro about gender inequality in guess who?) Online accessed, April 2, 2017. http://www.huffingtonpost.com/2012/11/21/jennifer-oconnell-hasbro-gu ess-who_n_2165482.html 45

The Metropolitan Museum of Art (2nd Century B.C.–4th Century A.D.). Twenty-sided die (icosahedron) with faces inscribed with Greek letters. Accession Number 10.130.1158. 59

The Metropolitan Museum of Art (30 B.C.–A.D. 364). Dice. Accession Number 10.130.1156. DOI: 10.1007/springerreference_12343 59

Tidball, J. (2011). Pacing gameplay, three-act structure just like God and Aristotle intended. In *The Kobold Guide to Board Game Design*, pp. 11–18, Open Design, Kirkland, WA. 31

Tresca, M. J. (2011a). *The Evolution of Fantasy Role-Playing Games*. McFarland and Company Publishers, Inc., NC. 16

Tresca, M. J. (2011b). *The Evolution of Fantasy Role-Playing Games*. McFarland and Company Publishers, Inc., NC. 60

Vera, A. H. and Simon, H. A. (1993). Situated action: A symbolic interpretation. *Cognitive Science*, 17(1):7–48. DOI: 10.1207/s15516709cog1701_2 12

Warner, E. and Warner, A. G. (1909). *Shahnama of Firdausi, trans*. Kegan Paul, Trench, Trubner and Co., London. 10

Wizards of the Coast (2003). *Dungeons and Dragons Player's Handbook: Core Rulebook*. Hasbro. 60

Yalom, M. (2004). *Birth of the Chess Queen: A History*. Pandora. 10

Yermolaieva, S. and Brown, J. A. (2017). Dice design deserves discourse. *Game and Puzzle Design*, 3(2):64–70. 12, 59, 74

Yianni, J. (2001). *Hive Pocket*. Gen42. 35

Young, R. M. and Cardona-Rivera, R. (2011). Approaching a player model of game story comprehension through affordance in interactive narrative. In *Workshops at the 7th Artificial Intelligence and Interactive Digital Entertainment Conference*. 12

Zhirosh, O., Brown, J. A., and Tickner, D. (2019). Democratizing faculty development—establishing a training program at a new computer science university in Russia. In *ASEE Annual Conference and Exposition, Conference Proceedings*, pages Accepted, Paper ID #25473, Tampa, FL. 16

Authors' Biographies

HAMNA ASLAM

Hamna Aslam received a B.Sc. (Hons.) in Computer Engineering from Bahauddin Zakariya University. She received an M.Sc. in Computer Engineering from the University of Engineering and Technology, Lahore, Pakistan. Presently, she is a Ph.D. student as well as an instructor at Innopolis University. She has published numerous peer-reviewed papers on the topic of human factors and game design.

JOSEPH ALEXANDER BROWN

Joseph Alexander Brown received a B.Sc. (Hons.) with first-class standing in Computer Science with a concentration in software engineering, and an M.Sc. in Computer Science from Brock University, St. Catharines, ON, Canada in 2007 and 2009, respectively. He received a Ph.D. in Computer Science from the University of Guelph in 2014.

He previously worked for Magna International Inc. as a Manufacturing Systems Analyst and as a visiting researcher at ITU Copenhagen in their Games Group. He is currently an Assistant Professor and Head of the Artificial Intelligence in Games Development Lab at Innopolis University in Innopolis, Republic of Tatarstan, Russia, and an Adjunct Professor of Computer Science at Brock University, St. Catharines, ON, Canada.

He is a Senior Member of the IEEE, a chair of the yearly Procedural Content Generation Jam (ProcJam), the proceedings chair for the IEEE 2013 Conference on Computational Intelligence in Games (CIG), and Vice Chair for the IEEE Committee on Games.

Printed in the United States
by Baker & Taylor Publisher Services